cv $2.00

ACID RAIN

ACID RAIN
KATHLYN GAY

Franklin Watts
New York | London | Toronto | Sydney
An Impact Book | 1983

Library of Congress Cataloging in Publication Data

Gay, Kathlyn.
Acid Rain.

(An impact book)
Bibliography: p.
Includes index.
Summary: Examines the direct and indirect effects of acid rain, how it is formed, how it is being studied, who are the culprits in the creation of this very damaging ecological problem, and what citizens and the government can do to alleviate it.
1. Acid rain—Juvenile literature.
[1. Acid rain] I. Title.
TD196.A25G39 1983   363.7'394   83-16650
ISBN 0-531-04682-6

Copyright © 1983 by Kathlyn Gay
All rights reserved
Printed in the United States of America
10 9 8

CONTENTS

Chapter 1
Riding with the Wind and Clouds
1

Chapter 2
Is the Sky Falling?
9

Chapter 3
Troubled Waters
15

Chapter 4
Effects on Trees and Crops
23

Chapter 5
Over Land and Sea
30

Chapter 6
Who Are the Culprits?
36

Chapter 7
Can Governments Legislate
Against Acid Rain?
44

Chapter 8
Weighing the Costs
55

Chapter 9
Detectives on Water,
Land, and in the Sky
65

Chapter 10
What Citizens Can Do
72

For Further Reading
81

Index
83

Acknowledgments

The author is grateful for the special help
and information provided by
Dr. Orie L. Loucks, research scientist at
Butler University, Indianapolis, Indiana;
Dr. Leon S. Dochinger, forestry sciences,
U.S.D.A. Forest Service, Delaware, Ohio;
Dr. Patricia M. Irving, ecologist,
Argonne National Laboratory, Argonne, Illinois;
Shell Perrigan, crop science department,
Oregon State University;
Harriett Stubbs, Project Director,
Acid Precipitation Awareness,
West St. Paul, Minnesota;
Anne Baker, executive secretary,
Lake Champlain Committee, Burlington, Vermont;
Dr. Anne LaBastille, ecological consultant,
Big Moose, New York;
and many others who are too numerous
to name but are not forgotten.

ACID RAIN

CHAPTER 1

RIDING WITH THE WIND AND CLOUDS

At the edge of a lake in the Adirondack Mountains of upstate New York, someone stops to peer into the clear blue water which no longer can sustain fish or other aquatic life. Inside a trailer—a mobile laboratory—parked near an airport in northern Wisconsin, chemists test the acidity of water samples that have been brought in by helicopter from lakes in the area. At isolated locations all across the northeastern United States, in Canadian provinces, in Scandinavian countries and in other parts of Europe, individual scientists and groups of researchers are testing and documenting the condition of surface waters—lakes and streams. Hundreds of these in the United States are so acidified that aquatic life is being threatened or has been destroyed.

Scientists in Canada, Germany, Scandinavia, and the United States have also reported decreasing growth rates for some species of trees in forest areas. In the United States alone vast numbers of trees are dying. For example, over half of the red spruce in the high elevations of the Green Mountains in Vermont have died since 1965. "It's a shocking statistic," says Dr. H.W. Vogelmann, chairman of the botany department at the University of Vermont. Dr. Vogelmann, who has been studying the problem since the late 1970s, adds that "the mortality is regional and is also occurring in the Adirondacks and in the White Mountains [of New Hampshire]."

Damage to other evergreens such as pine and fir has been reported in the eastern part of this continent and also in England, Germany, and Sweden.

In addition to the ecological problems, metals are corroding and world-famous sculptures and monuments are being eaten away. Already such historic structures as the Taj Mahal, the Parthenon, and the Statue of Liberty have been damaged.

During the past few years, automobile manufacturers have warned owners that the paint finishes on their new cars could be ruined by "a chemical fallout" which results in "blotchy ringlet-shaped discolorations, and small irregular dark spots etched into the paint surface." Paint damage on new cars happens more in the northeast than in any other part of the United States, but recently auto dealers in several midwestern towns have complained about similar problems. Their fresh-from-the-factory cars on sales lots have been marred with indelible spots. In one Indiana county eight new sheriff's patrol cars were marked by the "chemical fallout" during a week in August 1982.

The descriptions of damage and destruction fill the pages of hundreds of reports and include examples from many parts of the world. What strange malady has been affecting the global environment? It has been labeled *acid rain*. But more correctly it is the direct and indirect effects of acid deposition.

"WET" AND "DRY" DEPOSITION

The term *acid rain* is sometimes used very loosely to describe almost any type of air pollution that falls to the earth, but even in a technical sense the term is a "misnomer of sorts," as the Environmental Protection Agency (EPA)

Information Services points out. There is not only acid rain, but also acidic snow, sleet, hail, dew, fog, and frost. Such acid precipitation, though, is just one part of the general phenomenon called acid deposition, which includes both wet and dry acidic substances. These acidic substances are transferred from the atmosphere and deposited on exposed surfaces such as plants, soils, and bodies of water as well as buildings, statues, and other structures.

When scientists refer to *wet deposition*, they usually mean dissolved substances as well as particles removed from the atmosphere by rain, snow, and other precipitation forms.

In a similar way *dry deposition* is a term for the materials deposited (without precipitation) as well as the process of depositing substances from the atmosphere. The materials include dustfall or coarse particles which settle by gravitation, fine aerosols (such as smoke) which impact by air motion, and gases that are absorbed (taken in) or adsorbed (held on a surface).

With moisture, some pollutants in dry form can be transformed into acidic substances. Both wet and dry acidic deposits are part of a similar pollution problem, but wet deposition is the pollutant most people call acid rain. The problem is complicated further by gaseous ozone which is not acidic but may also be contributing to terrestrial damage.

MEASURING ACIDITY

How is the acidity of rain determined? Chemists use the pH (potential Hydrogen) measurement to express the acidity of water solution in terms of hydrogen ion concentration.

Ions are electrically charged particles. Positively charged ions are called *cations*, while *anions* are negatively charged ions. Hydrogen ions are positively charged, and a solution with more hydrogen ions than hydroxyl is acidic. A higher hydroxyl concentration means the substance is alkaline or basic.

The pH scale is usually numbered from 0 to 14. Seven is neutral, the value of "pure" distilled water (which is neither acidic nor basic). Values below 7 are acidic, while those above 7 are alkaline. Since the pH scale is logarithmic, each number represents a tenfold change in the concentration of hydrogen ions. A substance with a pH of 6 is ten times more acidic than "pure" distilled water at pH 7. A pH 5 measurement indicates one hundred times more acidity. Four is one thousand times and 3 is ten thousand times more acidic, and so on. In short, the lower the pH number, the more acidic the substance.

Acidity is not always harmful. Stomach fluids, which are essential for digestion, are highly acidic. Many foods and beverages are composed of organic acids. Apples and most carbonated soft drinks, for instance, have a pH of 3.0 while carrots and spinach are slightly acidic at the 5.0 and 5.4 pH values respectively. Some acidic substances are needed to release nutrients from soils for plant growth.

Theoretically "normal" or "clean" rainfall (without pollutants) has been defined as being slightly acidic due to carbon dioxide (CO_2) gas, which is a natural part of the atmosphere. Carbon dioxide dissolves in pure water to form a weak solution of carbonic acid having a pH of 5.6. However, both acidic and alkaline substances occur naturally in the atmosphere and enter normal rainfall in small quantities. Thus, unpolluted rainfall over a continental area has a mix of ions that may or may not produce a pH of about 5.6. Scientists have defined acid precipitation as rain or snow with pH values substantially below 5.6. When the pH in lakes and streams drops below 5, aquatic life is threatened. A pH of 4.7 is lethal to most species of fish, as will be described more fully in chapter 3.

INCREASING ACIDITY?

According to various studies rainfall in eastern North America (and most of Scandinavia) usually has a pH of 4 to

5, and acid rain with a pH of 3 to 4 is common in a few locations in western Pennsylvania and West Virginia. The pH has been measured in the 4.0 range in the megalopolis area stretching north and south of Denver, Colorado, along the Rocky Mountains. In Minnesota, along the U.S.–Canadian border in the wilderness called the Boundary Waters Canoe Area, increased levels of acidity in precipitation have been reported.

Rainfall is not usually strongly acidic in the western part of the United States, except for urban areas surrounding cities such as Seattle and San Francisco. In Los Angeles, California, acid fog with a pH of 3 has been recorded.

A number of scientists studying the acid rain phenomenon have concluded that the acidity of rain has increased since industrialization began about 1850 and has become more widespread since the middle of this century. However, the federal Interagency Task Force on Acid Precipitation pointed out in a recent report that

> there is considerable controversy about . . . trends in precipitation acidity over the last several decades. The acidity of precipitation has only been measured consistently for nearly two decades in one place in North America—the Hubbard Brook Experimental Forest in New Hampshire—and no marked trend in pH is evident in that record.

The report explains that it is difficult to detect trends in acid rain because different measuring techniques and methods of analysis have been used during the past thirty years. In addition, there have been no measurements of dry deposition. Yet even with such limitations, the overall data from various monitoring stations "suggest the possible spreading and intensification of acid rain," according to the report.

During the last few years, major monitoring networks have been organized in both the United States and Canada.

Some fifty monitoring stations have been set up by the Canadian Network for Sampling Precipitation Program (CANSAP). The U.S. National Atmospheric Deposition Program (NADP), organized in 1978, has established over one hundred stations. Monitoring efforts have also been undertaken by state and provincial governments in both countries.

Dr. Ellis B. Cowling, NADP chairman and a scientist at North Carolina State University who has been engaged in acid precipitation research since the early 1970s, says investigations show "long-term changes in the acidity of precipitation in North America . . . " which suggest that there is increasing and spreading acidity over the northeastern section of the United States and Canada.

SOURCES OF PRECURSORS

Why this increased acidity? Where does acid rain come from? Who or what is responsible?

The accusing finger points to the combustion of fossil fuels such as high-sulfur coal and oil and the smelting of sulfide ores. Fossil-fuel-burning and smelting operations send gaseous sulfur and nitrogen oxides (SOx and NOx) pouring into the air. It has long been recognized that sulfur oxides, particularly sulfur dioxide, and nitrogen oxides are major pollutants, but they are also what scientists term *precursors* or forerunners of acid rain. The gases transform into compounds called sulfates and nitrates. Although other substances may contribute to the formation of acid rain, it is known that infinitesimal particles of sulfur and nitrogen oxides react with moisture in the air and convert to sulfuric and nitric acids.

Sulfur dioxide and nitrogen oxides come from both human activities and natural events. Some acid-forming materials are released when volcanoes erupt, when organic matter decomposes, or when forest fires burn. Lightning and sea salt spray are other natural sources. However, in industrialized areas of the world, human activities account for a major portion of the acid rain precursors.

Each year in the United States almost 30 million metric tons of sulfur dioxide and over 15 million metric tons of nitrogen oxides pour into the atmosphere. The emissions come from electric utilities, large industries, smelters of nonferrous ores (copper, etc.), transportation systems, and heating systems. Scientists generally agree that utilities contribute 60 percent of the total sulfur emissions.

The exhausts of cars, trucks, planes, and other transportation sources send at least 9.7 million metric tons of nitrogen oxides into the atmosphere annually in the United States. The task force estimates that by the year 2000 total nitrogen oxides emissions in the United States from both mobile and stationary sources will be about 27 million tons, although the increase is expected to originate mainly from stationary sources.

Sulfuric and nitric acid concentrations of precipitation vary, depending on both time and place. Usually there is a higher nitric acid concentration in the western part of the United States due to transportation sources. In eastern North America (including Canadian provinces) sulfuric acid is predominant. By far, most of the sulfur pollution comes from industrial areas east of the Mississippi River. In Canada the smelting facilities at Sudbury, Ontario, are generating much of the sulfur dioxide. Sulfur emissions in the United States come primarily from the Ohio River Basin— an area where high-sulfur coal is burned in huge power plants and which includes Ohio, Illinois, Indiana, Pennsylvania, West Virginia, and Kentucky.

A RIDE WITH THE WIND

Where do the acid-forming materials go, once they get into the air? It is estimated that about half of the pollutants from power plants, industries, and mobile sources fall to the earth just a few miles away or within 18 to 30 miles (30 to 50 kms). But some sulfur and nitrogen oxides rise high into the atmosphere where winds will carry the compounds long distances, hundreds to perhaps thousands of kilometers from their point of origin. During transport, the pollutants react with water vapor and hours to several days—or even weeks—later may fall as acid precipitation.

Elevated chimneys, often called "tall stacks," on power plants and smelting facilities have been blamed for long-distance transport of sulfur compounds. It was once believed that effects from sulfur dioxide emissions could be abated through dilution, so the tall stacks—some 1,300 feet or more (400 m) high—were built to get rid of local pollution. But what goes up does come down—eventually—somewhere. The pollutants are just "airmailed" farther away.

Some scientists have concluded that prevailing winds, especially those from southwest to northeast, carry acid rain precursors from the Ohio River Valley to mountainous regions of the New England states. Canada has long claimed that at least 50 percent of the acid rain falling in its provinces is brought about by pollutants from the northern border states in the United States, while only 10 to 15 percent of the Canadian pollutants blow southward to form acid rain in the United States.

The impact of acid rain on the environment also depends on "the total amounts and spatial distribution of sulfur and nitrogen emissions. Substantial vertical and horizontal mixing of air pollutants from many different and geographically separate sources occurs during transport over hundreds of miles," Dr. Cowling points out.

The length of time acidic substances stay aloft is also a factor. Climate (the amount and frequency of rainfall, humidity, and solar radiation), geology (the type of bedrock and soils), the kinds of trees and other vegetation in an area where there is acid deposition are important aspects as well.

CHAPTER 2

IS THE SKY FALLING?

Since the mid-1970s countless magazine and newspaper articles, technical papers, scientific reports, brochures, booklets, and conference proceedings have been published on the subject of acid rain. Some reporters have described it as "Death [or Poison] from the Sky" or likened it to a "Reign of Terror" or an "Ecological Holocaust." One writer even claimed Chicken Little was right: "The sky *is* falling," alluding to effects just as disastrous from acid rain.

On the other hand, some commentators have declared that the acid rain issue has been blown out of proportion. There is little to be concerned about since "99.9 percent of all rain is acid in nature," as James Edwards, former Secretary of the Interior, grossly oversimplified the case. David Stockman, budget director for the Reagan administration, declared that if a few fish die (or are dying) because of acidity, the fish are not much of a loss when compared to the tremendous cost of controls to prevent the formation of acid rain in the first place.

Scientific discussions and political debates surrounding

air pollution controls have increased public awareness of the acid rain problem during the past decade. These aspects, along with effects on the environment, will be covered in the chapters ahead. Yet, some historical perspective is important when considering current controversies.

PAST STUDIES

Acid rain is not a recent phenomenon. Centuries before the term acid rain was used, it was recognized that industrial pollutants were a human health hazard and harmful to plants. In the mid-1600s English researchers proposed that factories with taller chimneys be built outside town boundaries. Industrial smoke could then be sent far from its source.

By 1727 another English scientist found acid and sulfur particles in the air and concluded that these same substances were components of rain and dew. Seven years later a Swedish researcher claimed that pollution from a smelter at Falun, Sweden, was "poisoning the air wide around" and destroying nearby plant life.

During the nineteenth century Robert Angus Smith, a British chemist, began analyzing the chemistry of rain in the city of Manchester and the surrounding English countryside. Writing in 1852, Smith described the changes in the rain chemistry from the outer fringes to the center of the city where there was air with "sulfuric acid or acid sulphates."

Smith was the first researcher to label acid rain, coining the term in a book titled *Air and Rain: The Beginnings of Chemical Climatology* (1872). He reported on studies conducted in England, Germany, and Scotland which showed that the chemistry of precipitation was affected by the same factors under study today, such as coal burning, decomposing organic matter, wind patterns, and the amount and frequency of precipitation. Smith also provided details on how to collect and analyze precipitation and described various damages to materials and plants due to acid rain.

MODERN RESEARCHERS

Although Smith's book was prophetic and recorded pioneering efforts in acid rain research, few modern scientists knew about his work. In 1981 the U.S. National Academy of Sciences published a report which has been widely circulated and includes the first analysis of Smith's findings. The analysis was developed by a Canadian scientist, Eville Gorham, who is now a professor of ecology at the University of Minnesota.

Dr. Gorham has been studying the environmental effects of acidic substances since the 1950s. His scientific reports have provided the basic understanding and foundation for further research on the consequences of acid deposition. Yet Gorham's work, like Smith's before him, was not recognized until the past few years.

The diversity of acid rain studies has made it difficult to coordinate findings in the scientific community. But that job was tackled by Svante Odén, a Swedish soil scientist. In 1967 and 1968 he brought together knowledge from the seemingly unrelated fields of agriculture, atmospheric chemistry, and limnology (the study of lakes and streams).

Reporting on the changes in acidity of precipitation, Dr. Odén demonstrated how pollutants can travel long distances, specifically from Central Europe and England to Scandinavia. He also explained how naturally acid soils could rapidly become more acid due to chemical processes triggered by acid deposition. Odén predicted other consequences such as fish kills in acidified lakes and streams, plant diseases, slower growth of trees in forests, and damages to various metal and stone structures.

Dr. Odén's reports caught the attention of the European press and the scientific community worldwide. Conferences were held to discuss and debate his ideas. During 1971 Dr. Odén lectured in the United States and Canada where he spurred public and scientific interest in acid rain. Since then a number of prestigious U.S. and Canadian scientists and various research groups have initiated studies on the causes and effects of acid deposition.

CANADIAN INTEREST

In North America Canadians were the first to investigate acid rain effects. There were some early studies of sulfur dioxide effects near smelting operations, but attention was focused on lakes after investigations were conducted by Harold Harvey of the University of Toronto.

Dr. Harvey, who is an expert in fisheries, attempted in 1966 to breed salmon in Lumsden Lake. This small but deep body of water, surrounded by mountains in a wilderness area of Ontario, should have been a near perfect spot for the four thousand salmon introduced. After breeding the salmon, Dr. Harvey hoped to stock other lakes with the sport fish. But a year later none of the salmon were left.

During an entire summer Dr. Harvey and his student team, using huge nets, searched for salmon in the lake. They found only dwarfed suckers, the fish salmon usually feed on. No other fish were caught, even though trout and perch apparently had been plentiful just a few years before.

Dr. Harvey and his research group spent two more summers studying other lakes in the area and found almost all fish had disappeared. During routine testing for acidity, they discovered the lakes had changed drastically. Government records for 1961 showed that Lumsden Lake, for example, had an acid concentration of 6.8 on the pH scale, quite safe for the fish population. But in 1971 the pH was 4.4 — much more acidic than it had been ten years before and deadly to the fish.

Similar conditions were found in many of the 150 lakes

that Dr. Harvey and his crew investigated through 1972. Eventually, after tests of rainfall and snow samples, he determined that sulfur dioxide, which transforms to sulfuric acid in water, could be coming from huge smelting operations near Sudbury, Ontario. He warned that the sulfates could be destroying the lakes, making them too acidic for fish or other aquatic life to survive.

U.S. SCIENTISTS

In the United States there had been some research on precipitation chemistry during the 1920s, and at mid-century a regional monitoring network was maintained for two years by scientists at State Agricultural Experiment Stations. From 1960 to 1966 the U.S. Health Service monitored precipitation chemistry nationwide. But it wasn't until 1975 that Dr. Ellis Cowling of North Carolina State University, along with other U.S. and Canadian scientists, began to develop a permanent monitoring network. Now sampling stations across the United States and Canada are collecting wet and dry deposits, and these substances are being analyzed for chemical changes.

Some American scientists are concentrating their efforts on the study of lakes in upstate New York. At Cornell University Dr. Dwight Webster had been collecting data on the dying fish populations in the lakes of the Adirondack Mountains. Then in 1976 Dr. Carl Schofield, a Cornell aquatic scientist, analyzed and published alarming reports showing the Adirondack fish kills were related to the acidification of the lakes. Other scientists at Cornell, notably Dr. Gene E. Likens and Charles Cogbill, have conducted a number of experimental studies to determine what biological damages and possible benefits have resulted from acid precipitation.

At the University of Virginia Dr. James N. Galloway developed techniques for collecting precipitation samples. Important acid rain research has also been underway at such institutions as the State University of New York (at Albany), Washington and Yale Universities, and the University of Vermont.

With the many investigations and debates on the real and apparent hazards of acid rain, Dr. Cowling of NCSU has pointed out that there is a need for "communication among biologists, hydrologists, atmospheric chemists, climatologists, regulatory strategists, and both industrial and political leaders in the United States and Canada." He calls for experts to focus their "collective knowledge on the development of more adequate plans for management of air quality and its effects on plants, animals, soils, surface waters, and materials."

While no responsible researcher is predicting that the sky will fall, many are sounding the alarm that the environment must be protected from acid rain effects. The quality of life depends on preserving the abundant production from fields, forests, lakes, and streams. If citizens as well as acid rain researchers are extremely well informed about the issues surrounding the pollution problem, it is possible that methods will be put into effect to manage our natural resources wisely.

CHAPTER 3

TROUBLED WATERS

In the New York resort area of the western Adirondack Mountains, Big Moose Lake is so acidified that swimmers have come out of the water with bloodshot eyes. Most of the fish have died, and other animal populations that once fed on the fish are diminishing. According to longtime residents in the area, there are not even swarms of insects over the lake that once kept the fish jumping.

Nearby at Woods Lake a similar situation exists because of the changes that have been taking place in the water chemistry. Brook trout, frogs, salamanders, otters, and mayflies are just some of the species associated with the aquatic ecosystem which had been plentiful but now barely survive or have disappeared.

Many other lakes in the Adirondacks are becoming what experts call "critically acidified"—the average pH which was about 6.8 in the 1930s now is below 5, and sportfish have been eliminated. There is concern that such a process is being repeated in the Boundary Waters Canoe Area of northern Minnesota and in similar wilderness areas in Wisconsin and Michigan. Surface waters are also in trouble in other parts of the United States, in Canada, and in European countries.

If you could take a look at the lakes and streams that have been affected by acid inputs, you might wonder what all the fuss is about. Some lakes appear "pure"—sparkling clear and blue. You might even be able to see to the bottom of a lake or stream. But the view is deceptive. Gone is the microscopic life that once "muddied" the surface. And at the bottom leaves and other plant debris may be piling up—the materials can accumulate for years without decomposing if the biological processes that normally break them down have been altered by acid deposition.

"BUFFERING" CAPACITY

Surface waters are not usually affected directly by acid "fallout." Instead, once deposited on the ground, acid-producing substances travel through watersheds (the region drained by a river) to lakes, and there many factors determine whether the water chemistry will be changed.

Some soils in the watershed may exchange nutrients, such as calcium and magnesium, for components of the acids—the nutrients may be leached from the soil and enter water bodies while acids stay in the soil. In loose soils with alkaline materials, the acids can percolate through slowly and be modified so that less acid ends up in lakes and streams. Sulfuric acid, for example, can be held in some soils for a period of time. The watershed may contain materials such as calcium carbonate, the compound found in limestone, which can buffer or neutralize the incoming acids.

Like soils, "virtually all lakes and streams have some acid-neutralizing capacities," says Dr. Loucks, a Butler University research scientist who has long been involved in studies of acid rain effects. He pointed out that "if the surrounding watershed contains little neutralizing material, natural alkalinity levels in lakes and streams might be quite low, making these aquatic resources highly sensitive to even low levels of acidic inputs."

North American lakes and streams which have suffered the most damage are carved from hard rock—primarily granite—which underlies most of the eastern United States and Canada and portions of the northern Great Lakes states. In such areas there are sometimes steep slopes and thin soils, so acid rainfall runs off before it can be neutralized. The soils may have few neutralizing/buffering materials and few elements to exchange or hold acidic substances.

ACIDIFIED WATERS

As surface waters receive acid inputs, they do not necessarily become acidic immediately. For some time it was assumed that alkaline or neutralizing substances in a lake were consumed as acidic material was added and that once the buffering capacity was "used up," the lake became acidic. But studies by the Electric Power Research Institute (EPRI) of three lakes in the Adirondacks show that lake acidification is a more complex process. Whether or not a lake becomes acidic depends on the "biogeochemistry of its entire catchment, terrestrial as well as aquatic," an EPRI report says. The EPRI studies have also shown that a lake is not always vulnerable to "acid fallout" if underlaid with granite bedrock. Even though such bedrock has "little neutralization potential," EPRI has found that deep mineral soil layers may buffer acid inputs.

When a lake or stream begins to lose its buffering capacity or ability to neutralize acids and the pH falls from 6 to 5, the aquatic ecosystem begins to show signs of stress. Amphibians such as frogs and salamanders are very sensitive to low pH and may be the first species affected. Many die or are deformed. Most shellfish also cannot live in low pH waters.

Algae are basic to the aquatic food chain, but as the pH

drops, acid-tolerant algae become widespread. Such algae are "not readily edible by zooplankton, the animals that link algae and smaller fish in the food chain. . . . Changes in algae community structure . . . may also alter zooplankton community structure," notes a report from the Office of Technology Assessment (OTA) which advises Congress on various technical matters being considered for legislation. As lakes acidify, there are fewer zooplankton, and the variety and size of the animals are reduced. Thus, less food is available for fish and other animals.

Studies of acidic water bodies in upstate New York, in Canada, and in Sweden show that acidification destroys aquatic plants. Then sphagnum mosses begin to cover the bottoms of lakes and rivers, trapping waterborne nutrients. These dense moss beds are not suitable for small organisms that live at the bottom of the lakes and streams or for fish spawning grounds.

When the pH of the water drops below 5, most fish do not survive. Those that do may be dwarfed or misformed. Fish kills are most often due to failures in reproductive systems. However, large numbers of mature fish also sometimes die—this most often during spring thaws, when acids, which have been stored in ice and snow, are released in the melt and rapidly run off into lakes and streams, causing major changes in the water chemistry. Such fish kills are the result of "acid shock."

Acidity itself is not the only cause of fish kills. The chemical reaction set off by acid rain may result in aluminum in the soil becoming soluble. Large amounts of the metal may be mobilized so that it is transferred to lakes where it is toxic to fish. Mercury may be released as well and in solution be converted to "highly toxic organic methyl mercury." EPA reports that "a clear correlation exists between the acid level of the lake and the mercury levels of its fish." Not only does the mercury kill fish, but it can be lethal to people who might eat contaminated fish that survive. However, there have been no known cases in the United States of mercury poisoning from freshwater fish.

HOW MANY ARE AFFECTED?

The number of lakes and streams being altered by acid inputs is startling. According to field studies hundreds of lakes and ponds in the Adirondack region of New York and thousands of surface waters in Canada are showing acid stress. Entire fish communities have been lost in many of the lakes and streams. Surface waters are also vulnerable in Michigan, northern Minnesota, and Wisconsin, and in parts of southern Appalachia and Florida and in large portions of Washington, Oregon, California, and Idaho.

In Colorado, on the so-called "wild side" or the western slopes of the Rocky Mountains, remote watersheds are receiving the deadly fallout from acid rain. Professor John Harte of the University of California, Berkeley, with a team of researchers, has been conducting a major study of the area for several years. Rain and snow samplings have registered "over and over again" pH levels as low as 3.6, Professor Harte reported in an *Audubon* Magazine interview.

The scientist and his team are now trying to learn where the acid deposition over the Rockies is coming from. Smelters and coal-burning power plants in western states and pollution from Los Angeles are suspected sources of acid rain in the high elevations of the Rocky Mountains. However, many studies must be completed before any conclusions can be drawn about acid rain sources.

An inventory of eastern U.S. lakes and streams considered sensitive to acid deposition was completed recently for Congress's Office of Technology Assessment. The OTA survey included a total of 17,000 lakes of 15 acres (6 hectares) or more in size and 117,000 miles (188,000 kms) of streams. These water bodies are in fourteen eastern regions where soils in the watersheds have little ability to neutralize acidic substances.

The acid in precipitation can move through these watersheds into lakes and streams. Also, the water may have low alkalinity levels and little ability to neutralize acids.

In any acid-sensitive region, alkalinity levels of water bodies vary due to differences in geology, soils, and runoff patterns. In the twenty-seven-state sampling area east of the Mississippi River, an estimated 9,400 lakes and 60,000 miles (97,000 kms) of streams were described as "currently sensitive to" or previously "altered by acidic inputs." Of that number some 3,000 lakes and 23,000 miles (37,000 kms) of streams are "acid-altered"—changes have taken place in aquatic plant and animal populations.

COUNTERMEASURES

Is anything being done to save some of our water resources? Control measures to slow down the shower of acid-producing compounds can help, and such regulations will be described later. But some more immediate steps have been taken to add neutralizing materials to a few lakes.

In New York, for example, biologists are treating half a dozen lakes with lime. The calcium compound in lime raises the alkalinity levels and the pH, thus saving fish populations. This can be a benefit in tourist areas where fishing is a popular recreational activity.

Similar efforts are taking place at Nelson Lake in Ontario, Canada, and at a number of locations in Sweden. The Swedish government has reportedly dumped more than 120,000 tons of lime in watersheds. This has kept the pH in lakes and streams at levels suitable for fish.

Liming, however, is not a simple task, and too much lime can cause an increase in toxic metals. Even with the right amount of lime, the neutralizing effect only lasts for a two- to three-year period. Also, it can be an expensive method for counteracting acid inputs. In some remote areas calcium compounds must be scattered over the lake surface by planes; airways may be the only access to high mountain regions. Sometimes tons of crushed limestone are dumped into a lake from a barge. It would take a tremendous amount of limestone to neutralize all the thousands of lakes and streams affected by acid deposition.

One other countermeasure is being studied by scientists. Several species of fish are apparently able to tolerate low pH levels. If such fish can be bred, they may survive in acidified waters. But experiments are still limited, and no conclusions can yet be made about stocking lakes with so-called "acid-tolerant" fish.

DRINKING WATER QUALITY

As previously described, toxic forms of aluminum and mercury can be dissolved and released from soils by acidic waters. Copper, lead, and asbestos are other toxic materials that may enter water systems.

Although studies of water quality show no direct connection between acid deposition and unsafe drinking water, harmful levels of lead and asbestos have been found in acidic tap waters of the eastern U.S. Scientists have not been able to determine the exact source of the toxic materials, but the OTA report warns: "Acidic precipitation can scavenge toxic materials from the *atmosphere* during rainfall events, leach them from soils and rocks as they pass through the *watershed*, or leach them from *pipes and conduits* used to distribute water to users." (OTA emphasis)

Drinking water from rural wells is more likely to be affected by acid deposition, especially in areas where soils cannot buffer acidic substances. There appears to be less danger from municipal water supplies since most are monitored regularly. If water quality depreciates, correction is fairly easy. Recent federal regulations have required corrosion control programs for public water systems. Any specific materials, such as lead, that might be present must be reported. In some areas where acid precipitation has affected water quality, treatments are required to prevent corrosion of conduits and plumbing materials.

High levels of asbestos have been found in the San Francisco Bay area drinking water. Asbestos is a natural part of the area's bedrock and has been linked with an above-average number of abdominal cancer cases. Natural supplies of asbestos also occur in rock formations all along the West Coast and eastern shore of the United States. "In addition, asbestos fibers are mixed in concentrations of 10 to 15 percent with cement to reinforce pipe used to distribute water supplies," the OTA report says.

If water corrodes the cement mixture, asbestos fibers may be released into the drinking water. But so far no data show a direct connection between acid rain and asbestos in drinking water.

Lead-contaminated drinking water is of real concern in some areas which may have acidified groundwaters. When the pH of water is below 6.5 and there are few neutralizing materials, lead may be corroded from water pipes and reach high levels of concentration. Erosion, street runoff, and leaded gasoline are some other factors contributing to lead in aquatic ecosystems. Some water samples have shown lead concentrations one hundred times higher than the standards set for "normal" water quality.

When toxic substances are released by acid deposition, it is not known how much the body actually takes in through drinking water. People are also exposed to poisonous materials in the workplace and may breathe in toxic substances from the air around them. There are still many uncertainties in regard to how human health is affected by acidified waters or, for that matter, overall acid deposition. Health effects—and costs—are intertwined with all other effects of acid rain.

CHAPTER 4

EFFECTS ON TREES AND CROPS

KILLER STALKS THE HIGH COUNTRY. The headline spread across the New England news section of the *Boston Globe* on a Sunday morning in May 1982. The chilling words were no exaggeration, even though there were no human victims. Some agent was at work killing off a large portion of the trees on the high slopes of the Green Mountains in Vermont.

Why should anyone be concerned about a bunch of fallen trees and barren stumps? Isn't that kind of scene familiar in forest areas where lumbering goes on? Drought, disease, or insects have often destroyed trees too. Or perhaps the slopes were being cleared for skiing, a popular sport in the New England mountains.

Apparently none of those factors has been causing the destruction of the 200- to 300-year-old trees, according to scientists from the University of Vermont who have been studying the area since 1965. At that time a data base on the mountain's ecosystem was put together. Records were made of the number and types of trees, the soil chemistry, moss cover, and climatic conditions. Scientists at the university believe they have the most detailed information available in the world about a natural area as background for forest research.

Dr. H. W. Vogelmann, chairman of the university's botany department, says that Camel's Hump (twin peaks east of Burlington, Vermont) used to be covered by dense green forest, but now the trees stand like skeletons or have toppled over, lifeless. Mosses have also decreased by 47 percent since 1965. There has been an increase in the depth of litter on the forest floor. Just as debris accumulates at the bottom of some acidic lakes, so the organic matter in the forest piles up. The material does not recycle as effectively as it once did.

Fog often shrouds the mountain area, and the precipitation can be as acid as vinegar. "Experiments done in our lab suggest that acid rain is a contributing factor in spruce die-back," Dr. Vogelmann says, but adds that "the forest environment is complex, and it is very difficult to show a clear cause and effect relationship."

Still the evidence points to a combination of acid precipitation and heavy metals in air pollution as the suspects in the "excessive mortality of red spruce which is a dominant tree in the high elevations of the Green Mountains," Dr. Vogelmann says.

OTHER FOREST STUDIES

The first studies of acid rain effects on forests were done in West Germany, where soil chemistry and forest growth were examined in an area downwind of the Ruhr industrial region. Changes in soil chemistry attributed to wet *and* dry acid "fallout" have seriously degraded beech and spruce forests there.

The West German researchers found that acid rain washes over several layers of leaves and branches before it gets to the ground. The foliage may be dusted with dry deposition—tiny particles of acid-producing substances.

Raindrops dissolve the particles and convert them to acids which may wash through or stay on the leaves.

If acidic substances on the leaves and from precipitation reach the forest floor, the acid input can be two to four times more potent than acid rain alone. This can slow the breakdown of decaying leaves and other matter which provide plant nutrients such as nitrogen. As the nitrogen increases in the soil, more acid is generated which can create additional stress for the trees.

In the West German state of Bavaria, some 13,500 acres (5,468 hectares) of evergreens may be doomed because of acid fallout. Reports of acid-threatened forests are also coming from Poland, Yugoslavia, and Czechoslovakia as well as England, France, and Switzerland.

Scientists have analyzed tree rings in the New Jersey Pine Barrens and have found a dramatic reduction in growth of the pine over the past twenty-five to thirty years. As in the Vermont study, many factors which might lead to decreased growth were investigated, but it appears that acid deposition has been the major culprit since about 1955. Soils in the region are poorly buffered, and the streams have become more acidic. New Jersey rainfall and stream pH levels have been about the same in recent years, averaging between pH 4 and 4.5. Although scientists stress that "no irrefutable evidence exists to relate the decline in growth rate to increasing precipitation acidity," there is a strong possibility that acid rain is a factor in the decline. Both the Vermont and New Jersey studies suggest that natural causes alone are not to blame for the changes in forest productivity.

Other forest areas which receive high levels of industrial and urban air pollutants and acid rain are also being studied. Eight major research sites—large forest areas—in the eastern United States and another unpolluted control site in the Midwest are part of a long-term investigation. The combined study should show whether forest productivity is affected by acid deposition. At each site cores will be taken from six species of trees to determine growth patterns. Researchers will also analyze soil chemistry, leaf nutrition, and wood density.

The forest products industry is especially interested in acid rain studies. Raw materials for the industry come from the forests, and even a slight reduction in productivity because of acidic deposition could result in major economic losses. In addition, manufacturing facilities of the industry emit sulfur and nitrogen oxides. If more strict control measures are required to reduce emissions and their adverse effects, the industry will have to pay a large share of such control costs.

SOME BENEFITS?

A few controversial studies claim that forests can actually benefit from sulfur and nitrogen deposits, particularly if soils are deficient in those nutrients. To determine whether a forest ecosystem is harmed or helped by acid deposition, scientists must consider the net effects of atmospheric input and the amount of nutrients in the soil.

> Most forests of the United States are nitrogen deficient, while a few (mainly in the northeast) are deficient in nutrients such as calcium and magnesium, and a few (mainly in the northwest) are deficient in sulfur [the OTA report explains]. Sulfur deposition throughout the eastern United States considerably exceeds forest sulfur requirements, whereas nitrogen deposition does not exceed forest nitrogen requirements and is probably of benefit to many forests.

Apparent fertilizing benefits from nitrogen inputs have been observed in studies of forests in Scandinavia. Yet calcium and magnesium are being leached from soils at a fast rate. Thus, benefits may be short-term. Some scientists believe that over a long period damages could result from overexposure to acid deposition.

EFFECTS ON AGRICULTURAL CROPS

When simulated acid rain (created in the laboratory) with a low pH has been sprayed on plants, detrimental effects may be seen but usually only at pH levels lower than those occurring now. These include damages to leaves and roots; slowdown of some biological processes such as nitrogen fixation; increased susceptibility to disease, insect injury, and cellular damage; and a decrease in the production of carbohydrates which can alter fruits, seeds, tubers, and roots—the parts of plants used for food.

Research has shown that acid rain can also affect the marketability of some crops. The quality, size, and appearance of the food products may be altered.

Still agricultural crops are not as likely to be affected by acid deposition as the trees in forests, and no direct injuries to crops grown in the field have been documented. One of the reasons for this is soil management. Farmers use fertilizers and lime on a regular basis, and such agricultural practices amend soils so that there are seldom any measurable effects of acid deposition.

Studies of simulated acid rain on certain crop plants have been conducted at the Agricultural Experiment Station, Oregon State University, in Corvallis. For two growing seasons crops were planted in pots and placed inside vinyl field chambers where they were sprayed with acidic solutions of pH 3.0, 3.5, or 4.0 and a *control rain* of pH 5.6. The *rain* was delivered through stainless steel nozzles for an hour and a half a day, three days per week for thirty weeks. Heat and air exchange were controlled by blowers. The same type of soil was used for all plants. Fertilization, irrigation, and pest and disease control were also regulated.

Of the twenty-eight species represented (including forage, grain, leaf, root, and tuber crops), in the Corvallis experiments, two-thirds were not affected. The yield of half the remaining crops increased, while the other half was reduced.

A number of scientists have pointed out that there have been conflicting and inconclusive results from research on possible effects to plants and soils caused by acid deposition. In the laboratory or greenhouse, some plants show direct effects if the simulated rain has a pH of 3.5 or less. But in the field, responses to acid rain depend on such factors as soil type, air quality, climate, differences in varieties of crops, and whether plants are covered to exclude natural rainfall.

OZONE EFFECTS

Scientists have long been aware that there was no interaction between various factors that influence plant productivity. Now it appears that interactions of acid precipitation and gaseous pollutants could have a damaging effect on plants. The experts are quick to point out, however, that there are still many uncertainties. Only the *potential* for negative impacts has been determined. According to the OTA report, regions most likely to suffer negative effects from a combination of gaseous pollutants and acid rain would be in the commercial forests of the southeastern coastal plains, agricultural and forest lands of the Mississippi River Valley, forests of the Appalachian Mountains, and agriculture and forest lands of the Ohio River Valley.

Ozone is one of the gaseous pollutants blamed for damages to plants. A secondary pollutant formed by chemical reactions of nitrogen oxides and hydrocarbons, ozone—like acid rain—cannot be controlled directly. Primary pollutants from which ozone is formed have to be reduced.

According to some estimates ozone may be responsible for up to 90 percent of the crop damage related to air pollution. Ozone has caused reduction in crop yield and quality of plants, and damage to leaves of many tree species. Yet it is not clear what concentrations of ozone cause detrimental effects and whether plant damages are due to ozone produced locally or far away.

A mix of ozone and acid deposition may also be damaging. It appears that crops are more likely to suffer from ozone effects if sulfur and nitrogen oxides are present in the atmosphere.

QUESTIONS RAISED

The possibility that acid rain can damage agricultural crops and forests is a major concern. There could be substantial economic losses. Thus researchers are now studying and experimenting to learn:

- which crops are affected by acid rain and in what regions of the United States;
- how acid precipitation affects plant products;
- how soil types, climate, air quality, and other factors relate to acid rain effects on plants;
- which crops might benefit from acidic deposition;
- whether crops are more susceptible to disease, insects, weeds, and injuries because of air pollutants;
- how acid rain affects the uptake and buildup of toxic metals and other toxic substances in crops;
- which agricultural practices help reduce harmful effects of acid deposition.

CHAPTER 5

OVER LAND AND SEA

Once acid-forming substances are released into the atmosphere, no one can be sure exactly how far the compounds will travel. But if sulfates and nitrates stay aloft for several days, it is possible for the chemicals to be carried across continents and oceans.

Apparently that is the situation in Europe. Industrial airborne pollutants from England and Germany move with the normal air flow and fall over parts of Sweden and Norway. Air masses travel hundreds of kilometers over the North Sea and along the Scandinavian coast, bringing acidic substances in the rain and snow. Acid deposits in thousands of Scandinavian lakes and streams have caused the death of fish and other aquatic life.

Government officials and scientists in Sweden and Norway have long protested the sulfur pollution from other European countries—particularly Great Britain, which has produced and exported the largest portion of acidic substances. Since westerly winds seemed to have carried pollutants from central Europe away from England, the British have not been too concerned about acid "fallout."

Recently, however, Great Britian has been hit by acid rain too. In a government funded study, scientists sampled rainfall at sixteen sites in Scotland. Over a large part of the country, rainfall is fifteen times more acidic than "normal." Researchers fear the rest of England may be receiving acid rainfall and are now studying possible effects throughout the country.

REMOTE AREAS POLLUTED

During the early 1980s, a number of reports were published describing the "dirty air" over the North Pole. Black soot is being carried to the Arctic from areas up to 5,000 kilometers away. The soot is not the same type of pollutant as acid rain, but scientists are trying to determine whether acidic substances travel as far and move along similar pathways as the soot.

Science Magazine recently reported that researchers with the U.S. National Oceanic and Atmospheric Administration (NOAA) have analyzed seventeen hundred samples of rainwater in Hawaii since 1974. Most of the samples were acidic with a range from pH 3.7 to about 5.7. The principal acidifying substance was sulfuric acid, but, as the report noted, scientists "could not find a sufficient source anywhere on the islands. . . ." Although Hawaii has active volcanoes, these were determined to be "insignificant sources" of sulfuric acid, and there appear to be no nearby industrial activities responsible for the pollutants.

Another study of a remote site is underway on Amsterdam Island in the central Indian Ocean. There rainfall averages about 4.5 to 5 on the pH scale, and rain samples show that sulfuric acid is also the acidifying substance.

Some researchers believe that sulfur compounds may be produced by biological processes in the sea, and these compounds can then be transformed in the atmosphere to sulfuric acid. But there is also a strong possibility that acidic pollutants from industrial centers are transported long distances to remote areas. One of the major difficulties in the acid rain studies has been tracking sulfur dioxide from its source to its final destination in the form of acid deposits.

ACROSS STATE AND NATIONAL BOUNDARIES

As you have read, the acid precursors in the United States travel in the atmosphere across state boundaries and also across northeastern borders to Canada. And some Canadian-produced acidic substances travel southward back over the border. A great deal of controversy has been generated by the pollutants which pass through the shared U.S.-Canadian airshed.

Many scientists believe sulfur compounds responsible for acid rain in the eastern part of North America come from the Ohio River Basin. High-sulfur coal is burned in large power plants in the area, and sulfur emissions from the Ohio Basin are then carried in the atmosphere to the northeast.

Recently Professor Donald Whitehead, a biologist at Indiana University (in Bloomington), completed a seven-year study which he contends links sulfur emissions from the Ohio Valley with acid rain over lakes, streams, and forests in New England and Canada. Professor Whitehead says more research is needed, "but there's no question that cutting outputs of sulfur from a broad Midwestern area would clearly benefit the Northeast's watersheds and lakes."

Dr. Walt Lyons, a meteorologist in Chicago who has studied thousands of satellite pictures of large masses of smog or air pollution, says these blobs (as the masses are called) can be followed from their origin to distant places. In a film Dr. Lyons produced, he demonstrates how an "urban plume" (smoke, gases, etc. from power plants and other sources) can travel downwind.

Using a computer simulation, the film shows what Dr. Lyons describes as a large-scale pollution episode over the Ohio Valley, which is being dominated by a stagnant high pressure system. The multitude of plumes travel for several days while undergoing photochemical reactions until a widespread haze, rich in sulfates and oxidants, envelops many states. The haze can drift many hundreds, if not a thousand or more miles off shore, and the entire episode is usually only ended by a cold front sweeping in from the west, bringing in an entirely new, clean air mass from the Pacific or from Canada.

Some smog or pollution blobs move in a clockwise direction from the Ohio Valley north through Illinois, Iowa, and Minnesota, Lyons says. Then the blobs turn toward the Great Lakes and up through Canada and drop down through New England and out into the ocean. He has seen such plumes about 1,500 miles (2,400 km) out into the Atlantic Ocean and speculates that a blob could even "end up in Europe."

The thick blobs are sucked into the clouds to make rain, Dr. Lyons theorizes. He points out in an EPA article that the satellite pictures show

> large masses of clouds with lots of little holes punched in them. For a long time we couldn't figure out why there were these holes, but then we realized that what had happened was rain—the smog had been absorbed into the raindrops and fallen out of the blob. We've verified this by comparing precipitation data taken from the ground with the satellite pictures of these holes. What we're seeing is acid rain actually being made.
>
> [According to Dr. Lyons] Pollution from any given area is not just a local problem. Due to shifting weather fronts, it can be a problem on a regional, national—or even a global scale.

ANOTHER VIEW

A somewhat different view has been expressed by Dr. Kenneth A. Rahn, a research scientist at the University of Rhode Island, who identified the source of soot over the Arctic. He believes the evidence that the Midwest is the single culprit in acid rain over the East is "quite indirect and circumstantial." In a statement issued in late 1981 regarding new studies of acid sources, Dr. Rahn noted that

> tracking a blob of haze is not hard proof [since, he believes] there is no way from photographs or visibility alone to determine how often the blob has renewed itself during transport. A blob may represent movement of certain meteorological conditions as much as it represents transport of mass.

Dr. Rahn also argues that

> just because the Midwest emits five times as much sulfur dioxide as the Northeast, it does not automatically follow that northeastern acidity should be dominantly midwestern in origin—the 600-mile (1,000-km) distance from the Midwest must negate much of this difference, especially in summer, when weaker circulation favors local pollution.

What about the view that the highest concentrations of sulfate aerosols (gases) are associated with the pathways of air masses from the Ohio River Valley? Dr. Rahn says that the relationship "implies little about whether a pollutant at the end of the trajectory [pathway] has actually traveled the entire distance."

While studying the Arctic atmosphere, Dr. Rahn and his associates at the University of Rhode Island developed a system of elemental tracers. Rahn sampled fine particles in the atmosphere to detect the elements vanadium (from oil) and manganese (from coal and ore). The ratio of manganese to vanadium should, according to Dr. Rahn's hypothesis, provide a "signature"—show the source of the pollutants—since more oil is burned in the East than in the Midwest.

Dr. Rahn's samplings have been limited, and he says he is not challenging "the basic picture of long-range transport." But he and his associates believe the Northeast has a "rich variety of sources and transport" of acidic substances.

Understanding of long-range transport in air pollution has increased a great deal over the past decade. But much of the information has come from studies of Scandinavia or the Arctic, where, as Dr. Rahn points out,

> . . . Sources of pollution are relatively easy to assign, because there is a single strong source far upwind and clean areas between it and the remote receptor region. But in the eastern United States there are no such clean areas; air traveling long distances is inevitably exposed to emissions all along its path. This makes it much more difficult to assign sources properly.

CHAPTER 6

WHO ARE THE CULPRITS?

Who will take the blame for acid rain? How can the pollution problem be controlled? Those questions have created countless debates in government and industry and among environmentalists and consumers.

On one side of the controversy are those who believe the federal government should pass an amendment to the Clean Air Act which would require reduction of sulfur dioxide emissions—more than presently allowed. This would affect many industries in the Midwest, particularly coal-mining operations and electric utilities which burn high-sulfur coal. Those against more government regulations say electric rates would go up considerably due to additional costs to control pollutants.

NO PROOF, SOME SAY

Senator Richard Lugar from Indiana defended midwestern industries recently. During an interview with the *Indianapolis News*, he expressed "enormous doubt that Midwest smokestacks are to blame for the content of rain falling hundreds of miles away . . . ." He felt "the real culprits" could be the "millions of cars in New York and the rest of New England . . . and local industries in those states. . . ."

Public Service of Indiana, the largest electric utility in the state, has presented similar views in a number of published statements to consumers. Basically, as with most utility and coal industries' stands on acid rain, PSI believes corrective measures are premature and is calling for much more study on the problem.

> The popular belief is that rain in the eastern United States and Canada is becoming increasingly acidic, that it's harmful to vegetation and wildlife and that coal-burning utilities in the Midwest . . . are the primary cause. . . . But the fact is that none of these three basic premises has been proven. To the contrary, there is mounting evidence that all three may be incorrect [PSI says].

The Coalition for Environmental Energy Balance (CEEB) based in Columbus, Ohio, echoes the arguments, saying the causes of acidification are not clear. "Statistics gathered from six midwestern states showed that coal consumption increased by an average of 12.5 percent from 1975 to 1980. And yet during that same five-year period, sulfur dioxide emissions decreased by 15.9 percent. . . . " CEEB states.

Public Service Indiana uses the same statistics to conclude: ". . . there is no corresponding downward trend in rainfall acidity. A 'cause and effect' relationship between sulfur dioxide emissions and acidity does not appear to exist."

Dr. Orie L. Loucks, research scientist at Butler University, has often countered such arguments. Loucks explained that the "acidity of rainfall has stayed about the same because of recent increases in nitric acid content . . . acidity that is presently of less quantitative significance." He also emphasized that "sulfur dioxide emissions abatement *is being matched by reduction in deposition of sulfate,* the principal acidifying agent."

Although scientists cannot prove without a doubt all of the mechanisms involved, there is scientific consensus that sulfur and nitrogen oxides are the predominant acidifying species in the atmosphere. The situation is very similar to links between smoking and cancer. Medical experts cannot prove by direct observation that a tobacco solution on the human lung will turn it cancerous. But statistics show that a larger number of people who smoke cigarettes for twenty years or more develop lung cancer than those who never smoke.

ACID CALLED "SCARE WORD"

Along with other charges and countercharges in the acid rain debate are claims that the media have sensationalized issues. Panic and fear have been created by news stories, say power and coal company spokespersons. The acid rain phenomenon is "little understood but, in recent years, much publicized," says the Columbus, Ohio-based American Electric Power System, which has coal-burning plants in seven states.

Just the term *acid rain* sounds evil, and the words create "menacing visions," some claim. "In our vocabulary 'acid' often implies strong actions, such as 'acid test,' or even unpleasantness, such as 'acid tongued,' " a representative for Edison Electric Institute, an association of electric companies, stated in response to the acid rain controversy.

Another EEI representative, Director of Legal Affairs William Megonnell, recently wrote in the *Environmental Forum* that the "street name" (acid rain) is a misleading, unscientific, and prejudicial epithet that implies guilt even before that has been established."

Some of the materials published by power companies or research firms hired by utilities go to great lengths to explain that "acid" isn't necessarily a "bad" four-letter word. Diagrams and illustrations are designed to show that acids are part of everyday life. The underlying message seems to be: "Have no fear 'cause acid's here!"

Certainly acidity is a natural part of the environment and essential for many biological processes. But long prefaces about the benefits of acid appear to be attempts to play down the possible harmful impacts of acid deposition on the environment. In fact, General Motors released a report in 1981 which claimed nitrogen oxides (the precursors of nitric acid) may actually combat air pollution. GM scientists say the more nitrogen oxides in the atmosphere, the less ozone there would be to pollute. Of course, it would be an economic advantage for GM to produce vehicles without strict emission controls. But consumers would probably save no more than sixty dollars on the price of a new car. Scientists not aligned with the industry believe it is foolhardy to try to reduce some pollutants by increasing others. They believe it makes more sense to cut back on emissions of all pollutants.

That logic is also attacked by utility representatives. The costs for added controls (see chapter 8) would be far too high, power companies say. According to some utility reports, electric rates would increase up to 50 percent in some areas of the Midwest.

Again there are arguments on the other side. The figures on cost increases, some believe, are based on contrived "worst case" situations in which older (and higher-polluting) plants are closed and companies write off their investments over a few years instead of over the long term. Several independent studies show electric costs in the Midwest would rise only 7 to 10 percent.

IS COAL THE REAL CULPRIT?

Oil—not coal—may be contributing more to acid rain pollution than coal burning. That is the opinion of PEDco, a research firm in Cincinnati which conducted a study for the Department of Energy. Based on fuel use patterns in Europe, Japan, and the United States, the study was made to determine whether acid rain could be related to changes in fuel usage.

In the industrial areas studied, fuels from petroleum are "more widely used than coal," and it is in these areas that acid rain occurs, the report states. To explain why oil burning could be responsible for acid rain precursors, the study points out that in Japan control measures have been established for boilers using residual oil; as a result sulfate concentrations have been reduced. The report also states that acid concentrations in rainfall have decreased in The Netherlands because of "the reduction of emissions from residual-oil-fired boilers."

Such findings are related to an increase in oil burning in the northeastern United States, particularly by electric utilities in the area. "Although sulfur content of the oil burned has been reduced, fine particulates that may function as catalysts in the formation of sulfates are being discharged. Also, these utility boilers appear to be especially productive of primary sulfates, which can participate directly in acidification of rain," the study concludes.

Nevertheless, the Research Council of the National Academy of Sciences issued its "Atmosphere-Biosphere Report" in late 1981 that points to coal-burning power plant emissions as major sources of acid-forming substances. The highly respected research panel also has recommended that sulfur dioxide emissions from power plants be reduced by 50 percent. Even though critics claim that there is no direct evidence linking emissions from power plants to acid rain, the research council finds their role "overwhelming."

SIMILAR LAKES, DIFFERENT ACIDITY

One of the arguments often used by spokespeople for utility companies concerns a study of three neighboring lakes in the New York Adirondacks which have different acid concentrations. The lakes are under investigation by scientists who received grants from the Ecological Studies Program of the Electric Power Research Institute (EPRI), sponsored by the electric power industry. This research effort, which has received high praise from the scientific community, has shown that one of the three lakes—Woods Lake—has an average pH of 4.5. Only a few acid-tolerant animal and plant species survive in the lake.

Fish still thrive at Sagamore Lake with an average pH of 5.5 and at Panther Lake with a pH of 7.0. Since the lakes are only about 20 miles (32 kms) apart and receive the same kinds and amounts of acid deposition, some claim this is evidence that acid rain has little to do with acidifying lakes. Other factors are at work, power company representatives insist.

Yet Dr. Robert W. Brockson, program manager for EPRI's ecological studies, explains why each lake has a different chemistry:

> We have found that the lakes are located on surficial deposits brought in by glaciers. The surficial tills are not the weathering by-products of the underlying granitic bedrock, but rather of material containing a larger percentage of buffer-producing minerals. The acid lake receives more surface and shallow subsurface flows from its watershed. The most alkaline lake receives a large fraction of its inflow from deep seepage, that is, groundwater that was exposed to a greater quanitity of buffers. . . .

In short, some Adirondack lakes have "buffering capacity" and do not become acidic even though they receive pH 4.0 rain.

CANADIAN CONSPIRACY CHARGED

"The Canadians Are Coming"—to compete with U.S. utilities; Canada wants to export electricity from its power companies to the United States. That is one of the theories presented by those who are critical of Canadian attempts to publicize harmful acid rain effects. Some U.S. industry and government officials also believe Canada is pushing for acid rain controls in the U.S. while the Canadian government does little about cutting back its own acid-producing emissions.

Michael Perley of the Canadian Coalition on Acid Rain recently wrote in the *Environmental Forum* that standards for air quality are "as strict as—and in some cases stricter than—standards under the U.S. Clean Air Act." Perley also points out that "Canadian sulfur dioxide emissions have declined by 25 percent during the past ten years, while U. S. emissions have remained relatively constant."

While speaking to a U.S. congressional caucus in early 1982, W. Charles Ferguson, director of Canadian government programs for INCO, Ltd., a huge smelting operation in Ontario, acknowledged that Americans "hear charges of loose or nonexistent Canadian legislation and hear repeatedly of the 'unlimited' emissions from INCO's Sudbury smelter." He said it appears that Canada is not taking any action while expecting America to adopt stricter emission controls.

> As the owner and operator of the INCO Sudbury smelter in Ontario, we realize we have become somewhat of a symbol of Canada's environmental control efforts [he said]. This symbolism is derived from the fact that the INCO Sudbury smelter is the largest point source of sulfur dioxide emissions in North America and emissions are vented from a 1,250-foot (380 m) tall chimney, the world's tallest. . . . [But] let me tell you that our emissions . . . have been limited by regulations of the Ontario government since 1970 . . . emissions from our Sudbury operations have been reduced to about one-third of what they were in

the peak years of the 1960s. In 1980, the Ontario and Canadian Environment Ministries established a task force to determine the lowest possible level of emissions that INCO might achieve. The findings of the task force will, undoubtedly, provide the basis for future orders to reduce emissions still further.

So who or what is to blame for acid rain effects? The controversy goes on—especially in relationships with Canada. The acid rain issue in the United States has become politicized, Canadian officials charge. They believe the environment will be sacrificed in favor of coal and power company interests. Many Canadians are concerned that the United States will allow sulfur emissions to stay at the same level in spite of the fact that scientific evidence suggests acid-forming pollutants from the United States are causing a lot of damage in Canada as well as in the United States.

CHAPTER 7

CAN GOVERNMENTS LEGISLATE AGAINST ACID RAIN?

President Jimmy Carter, in 1979, emphasized that acid rain was a global environmental problem and proposed a tenyear research plan to be managed by an Acid Rain Coordinating Committee. Members of the committee included representatives from various federal agencies and departments.

When Ronald Reagan was elected president, his new administration replaced the committee with the Interagency Task Force on Acid Precipitation, established under a Title VII program of the Energy Security Act of 1980 (PL 96-294). The group is led by directors of the Department of Agriculture, the Environmental Protection Agency, and the National Oceanic and Atmospheric Administration.

One representative from each of the departments of Commerce, Energy, Health and Human Services, Interior, and State are represented on the task force. Other representatives are from the Council on Environmental Quality, the National Science Foundation, and the Tennessee Valley Authority. The directors of the four national laboratories— Argonne, Brookhaven, Oak Ridge, and Pacific Northwest—

and four presidential appointees are included in the task force as well.

Although the task force has no regulatory powers, it can make recommendations regarding national policies on acid rain. However, a primary responsibility is carrying out a National Acid Precipitation Plan for research over a ten-year period (to 1992).

CLEAN AIR ACT

The Clean Air Act was passed in 1970 to set standards for local "ambient" air—the atmosphere near industries which emit air pollutants. Under the law individual states are responsible for enforcing clean air regulations and setting limits for local emissions. Thus, air quality standards can vary from state to state.

To comply with the clean air laws, power companies and smelters built tall chimneys to get rid of emissions. Since 1970 at least 175 smoke stacks over 500 feet (152 m) high have been built. Although local air quality standards have been met in some cases, the tall stacks merely send pollutants farther away.

In late 1970 the Environmental Protection Agency was established to enforce laws designed to protect the environment, including the clean air laws. By 1977 amendments to the Clean Air Act provided that EPA limit tall stacks. However, such measures have not resulted in much of a decrease in total emissions.

The 1977 amendments also gave EPA more authority to enforce state pollution control laws—if a state petitions EPA, charging that another state is polluting across borders with its emissions. In addition, EPA has the responsibility of carrying out regulations for new coal-fired power plants to use the best available control technology.

There is a problem, though, with older power plants which are not controlled by the regulations. It is estimated that old power plants now in operation will account for 73 to 75 percent of the sulfur oxides in the atmosphere through the mid-1990s and possibly to the year 2000. Also, there has been an increase in the use of coal which will keep sulfur dioxide levels about the same for some time.

ACID RAIN BILLS

Members of Congress have introduced several bills to amend the Clean Air Act as a means of dealing with the acid rain problem, but the proposed bills have created heated controversy. One of the reasons for the debates is that congressional representatives from the Northeast want controls to reduce sulfur dioxide emissions from the Midwest. As Senator Robert Stafford from Vermont was quoted in a recent issue of *Science* Magazine: "I don't think my neighbor should be permitted to throw garbage on my lawn, and say it's too expensive for him to hire a garbage man, and it's cheaper for him to simply continue throwing it on my lawn." Of course the senator was referring to atmospheric "garbage"—air pollutants from neighbors in the Midwest being dumped over the Northeast.

With that view in mind, Senator Stafford has cosponsored a bill, introduced by George Mitchell of Maine, that would require more stringent controls of emissions from the Midwest. Other members of Congress have also taken up the cry and have introduced, cosponsored, or supported similar proposed legislation to reduce Midwest sulfur dioxide emissions. The bills vary in the amount of reductions, the extent of the region deemed responsible for acid deposition, and the timetable for accomplishing controls. Generally, though, legislators seek to reduce sulfur dioxide emissions by 35 to 40 percent in states east of the Mississippi River.

On the other side of the dispute, midwestern representatives do not believe it is right to fix blame until more is known about acidic pollutants—how they are formed and transported and what effects are produced. That is also the position of power company officials who claim that there is not enough evidence to conclude that emission reductions would cut back on the acidity of precipitation. Many feel that control legislation would be "premature" and cost companies billions of dollars to comply.

According to Kathleen Bennett, an assistant administrator in EPA, the Reagan administration agrees and does "not favor an immediate imposition of an emission control program."Such a policy is based on "the fact that the Clean Air Act of 1970, with its 1977 Amendments, is already addressing all . . . pollutants of concern," Ms. Bennett said. She acknowledged that there is "an apparent difficulty" in regard to the Clean Air Act since it deals with local air pollution and "does not directly deal with secondary products of pollution that may be experienced far from the source. Given the vigorous prevention and control program already in place. . . the question . . . is whether an *additional* program is necessary. . . . "

Ms. Bennett has also expressed many times the White House position that more research is needed before a national policy on acid rain can be formulated. In spite of such views, a Senate Environment Committee voted during the summer of 1982 in favor of more strict emission controls. After many briefings on acid rain effects by highly qualified scientists, the fifteen-member committee unanimously approved an amendment to the Clean Air Act which would require reduction of the 20 million tons of sulfur dioxide emissions allowed per year since 1980. If the amendment is passed by Congress, electric utilities in the East and Midwest would be required to reduce sulfur dioxide emissions to 12 million tons per year by 1994 as a means of controlling acid rain.

CANADIAN–U.S. TREATY PROBLEMS

Canadians have expressed great dissatisfaction with what many officials believe are U.S. government stalling tactics in regard to acid rain controls. In August 1980 Canada and the United States signed a Memorandum of Intent. The two countries agreed to start immediately to put together data on pollution that crosses their shared boundaries. Joint technical and scientific work groups were set up to analyze the assembled information. This was to be a basis for Canada and the United States to negotiate a transboundary air pollution treaty.

The joint work groups have prepared two drafts of documents, but a third draft, which was supposed to be the final agreement, has been held up. In one part of the text, Canadians want to specify a maximum amount of pollutants that can be tolerated in the atmosphere. The tolerable amount of "loading" would result in an estimated 50 to 80 percent decrease in sulfur dioxide emissions. But U.S. members of the working groups who were appointed by the Reagan administration say there is no scientific evidence supporting a "loading limit." Yet the joint Canadian–U.S. groups had earlier agreed there was a scientific basis for the sulfur dioxide reduction.

Critics have charged that the U.S. scientists who are "strongly concerned about the environment" have not been popular. These scientists have been replaced in leadership positions by those who are more in line with the president's wait-and-see policy.

One of the replaced scientists, Dr. Gary Glass, a leader in aquatic science, has been described by his colleagues as a "thorough and comprehensive" researcher. He has also been called "a tower of strength" in aquatic research. But his job as cochair of the national aquatic study team has now been taken over by a crop specialist.

When interviewed by a news reporter, Dr. Glass said "everything [in acid rain research] is being structured to slow the whole process down as though we are ignorant. . . . " He charges that people are kept in power

> who do not have the expertise to see what is wrong. If they do not understand the problem there is no problem. . . . When you end up

showing them there is a problem, they have a requirement to do something about it. But it is a policy decision that more research is needed. So we put in policy people and fire the technical experts. Then we can make good on our promise that more research is needed before we act.

ADDED TENSIONS

Other administration tactics support the claim that the acid rain issues have been stalled and politicized. Independent scientists representing the National Academy of Sciences, who reviewed the quality of the technical papers for the proposed Canadian-U.S. treaty, advised a 50 percent reduction of sulfur dioxide emissions. The NAS report was not accepted by the U.S. government, but it was approved by Canada.

In late 1982 the United States called for acid rain reports to be reviewed by a committee appointed by the President's Office of Science and Technology. Canadians believe there should be a joint review with scientists from Canada and other countries as well as those in the U.S. The purpose of a review is to present research to experts in the field so that mistakes or conclusions based on insufficient or poor evidence can be pointed out.

Because of U.S. actions Canadian officials suspect that the United States is not negotiating an acid rain treaty in good faith. In addition, they point to the fact that the Reagan administration has supported relaxation of controls on sulfur dioxide emissions and that EPA is in favor of lower standards for emissions from auto engines. Canada believes these policies have undermined the Memorandum of Intent which calls for both Canada and the United States to "promote vigorous enforcement of existing laws and regulations" which limit emissions. As one Canadian environmental officer put it: "It's hard to arrive at a common solution to common problems unless you proceed jointly."

PROPAGANDA IN THE RAIN?

To put further strain on Canadian-U.S. relations, the U.S. Justice Department announced in late February 1983 that three Canadian films—one on nuclear war and two on acid rain—had been classified as "political propaganda." The film about the consequences of nuclear war, which two months later won an Academy Award, and the two acid rain documentaries—*Acid from Heaven* and *Acid Rain: Requiem or Recovery?*—were produced by the National Film Board of Canada.

Since the Canadian film board has to be registered in the United States as a foreign agent, a requirement of the 1983 Foreign Agents Registration Act, a list of all films the board produces and circulates in the United States has to be sent on a routine basis to the Justice Department. Of the listed films currently circulated in the United States, five were viewed by the department and two of those were cleared early in 1983. But the nuclear war film and the two films on the hazards of acid rain were labeled "political propaganda" and had to bear disclaimers that their contents had not been approved by the U.S. government. The Justice Department also requires that the Canadian film board submit a list of individuals and groups who request the film for viewing.

The Canadian government has protested the action and Canada's ambassador to the United States has asked the U.S. Justice Department to reverse its decision. Many Canadian officials have been angered by the classification of the film. John Roberts, Canadian Environment Minister, called the propaganda labels an "extraordinary interference with freedom of speech" and an attempt "to stifle the acid rain debate."

A number of U.S. senators and representatives have accused the Reagan administration of trying to suppress information on such politically sensitive subjects as acid rain and nuclear arms. And the American Civil Liberties Union has charged that the Justice Department action is "blatantly unconstitutional." On March 9, 1983, the ACLU, along with the state of New York, the New York Library Association, and various film distributors and groups of film viewers, filed a suit in U.S. District Court requesting that the Justice Department drop its control on the distributions of the three Canadian films.

Defending the Justice Department's actions, a spokesperson for the department claims the propaganda labels on the films are simply "to notify viewers that the material is being disseminated by a foreign government . . . not unlike the disclosures that are required on almost all political advertisements and commercials . . ." However, such disclosures are not required for *all* films of foreign governments.

ANOTHER POLLUTION AGREEMENT

The transboundary acid rain problem is often compared to a situation which existed before the Great Lakes Water Quality Agreement between the United States and Canada was signed in 1972. During the 1960s there were debates about the role of phosphorus in the clogging of lakes and streams with algae that destroyed fish and other life. Both Lake Ontario and Lake Erie of the Upper Great Lakes were affected, and experts could not agree on how much of a decrease in phosphorus "loading" would be needed to correct pollution problems. Some thought nitrogen was to blame, not phosphorus.

Dr. Ellis Cowling of North Carolina State University pointed out to a congressional committee that in the late sixties and early seventies industry was calling for more research on the effects of phosphates in the waters. At the same time environmentalists warned that there could be "irreversible damage" to the lakes.

> Finally a political decision was reached—a plan should be developed. More debates were held. A theoretical model was used to predict that a 50%

decrease in [phosphorus] loading might be sufficient. Under conditions of substantial uncertainty a management plan was finally implemented. After some time the Lakes began to improve. The theoretical estimate was too low, but with some further adjustments the plan worked and the Lakes are now on the road to recovery [Dr. Cowling stated].

INTERNATIONAL COOPERATION

According to experts on international law or legal treaties, the problems of transboundary air pollution cannot be solved by present international legal documents. International agreements on acid rain cannot be enforced because there are no means of policing or forcing control of emissions that produce acid deposition. Instead, cooperation between nations seems to be the key to any long-term solutions for global environmental problems caused by acid rain.

To that end countries which are members of the United Nations Economic Commission for Europe (the United States included) signed a Convention on Transboundary Air Pollution in 1979. The agreement calls for nations to cooperate in acid rain research and monitoring. Another international group, the Organization for Economic Cooperation and Development, has also recommended various ways countries can try to prevent transboundary pollution. But there are no methods for enforcing either decree.

In June 1982 a Conference on Acidification of the Environment was held in Stockholm with representatives from a number of European nations, Canada, and the United States. The purpose of the convention was to present scientific reports on the ecological effects of acid deposition and possible strategies to control emissions of sulfur and nitrogen oxides.

The conference noted the importance of the 1979 resolution on Transboundary Air Pollution and urged those nations which had not signed the agreement to do so. The 1979 agreement was called "a clear recognition that acid deposition from air pollution, including long-range transboundary air pollution, is one of the major environmental problems, requiring policies for further urgent action at the national level and concerted international efforts."

Based on reports from experts, the conference agreed on a number of concrete actions. Among them was "the establishment and implementation of the concerted programmes for the reduction of sulphur emissions" which the conference considered "a matter of urgency." Reducing nitrogen oxide emissions was another action the conference said "should be taken as soon as possible." The conference also supported research and development of advanced control technologies, energy conservation measures, and standardized and improved sampling and measurement of acid deposition.

Development of continuing public dialogue on acid rain issues was stressed by the conference. The international body recognized the value of nongovernmental organizations in debate/discussion of acid rain issues. They encouraged presentation of scientific information to the public "in an appropriate form."

STATE ACTIONS

Meanwhile, at the local level some states have adopted policies that are intended to curb acid fallout. In July 1982, for example, New York State and Quebec officials signed an agreement to coordinate their efforts to combat acid precipitation. New York and Quebec are carrying out joint studies, exchanging information, and standardizing methods used to analyze acid deposition. The two parties are also jointly preparing "a course of action to influence, on a national basis, decisions in favor of emission reductions for acid-causing pollutants," as outlined in the agreement.

Across the land in California there are attempts to counteract acid rain. California's problems are different from other states' in that the acid rain is believed to be caused mainly by nitrogen oxides (from cars and heavy industry) rather than from sulfur compounds thought to be primarily responsible for acid rain in the eastern United States. A California Assembly Select Committee on Acid Rain found that acid levels over much of the state averaged ten to a hundred times the normal value. In metropolitan areas acidity has measured fifty to one thousand times higher than the normal pH.

Long-range transport of acidic substances does not appear to be a problem in California. Most of the acid deposition comes from within the state. To combat the pollution California has strict standards and enforcement for control of auto engine emissions. The state effort is more stringent even than the existing federal program.

Other states, notably Minnesota and Wisconsin, have established acid rain programs also. But the problem does not always stay within state or even national boundaries, as has been noted. So there is little doubt that there must be more cooperative efforts among states and among nations if there is to be any improvement in pollution problems caused by acid fallout.

CHAPTER 8

WEIGHING THE COSTS

In almost every discussion about acid rain, the costs to the environment and human health are pitted against the costs for emission controls—or vice versa. The basic question is: Will there be enough benefits from pollution abatement to justify the huge expenses to cut back on emission of sulfur and nitrogen oxides?

There are no clear-cut answers, but weighing one type of cost (losses from damages) with another (expenses for controls) is a major issue in the acid rain debate. The issue becomes very complex if you try to determine the worth of wildlife areas, various artistic works and monuments, and the quality of human life affected by acid deposition or by higher electrical costs.

ESTIMATING DAMAGES

When it comes to environmental, materials, and health damages from acid rain, some estimates of losses have been made. According to the National Academy of Science, damages to agriculture, forests, lakes, water supply systems, materials, and health during a single year in the northeastern United States totaled about $5 billion. Another estimate for the Adirondacks region alone placed damages due to acid fallout at about $250 to $500 million per year. The tourist industry and recreational fishing suffer the most. Tourism has already declined because sport fishing in the area is so poor.

Recently Robert Flacke, commissioner of New York's Department of Environmental Conservation, said that loss estimates in the state have not yet taken into account possible damages to New York's $2 billion small-crop agricultural base. "We are taking a closer look at the true costs of acid rain damage in New York State," he emphasized, adding that

> I, for one, have not forgotten the statement by David Stockman . . . when he ridiculed the value of fish life in the Adirondack lakes. This study will show that economic damage from acid rain goes beyond fish alone. I am sure when the true cost of this damage is shown, no one—including Mr. Stockman—will volunteer to pay the bill.

Canadians are especially concerned about economic losses. Their billion-dollar fishing industry is threatened because of decreasing fish populations, and forest damages—some of which is thought to result from acidic deposition—can affect jobs. Over 10 percent of the work force in Canada is employed in forestry, and the industry itself is valued at about $4 billion annually.

Wilderness areas and national parks in both Canada and the United States are particularly sensitive to acid deposition. Losses of natural resources in such areas cannot be measured in monetary terms. How, for example, can anyone put a price tag on fishing pleasures no longer enjoyed? And what about possible forest, stream, and wildlife losses in such tourist attractions as Yosemite in California, the

Rocky Mountain National Park in Colorado, and the Great Smoky Mountains in North Carolina? Can natural beauty and vacation memories be valued in dollar amounts?

HEALTH COSTS

Along the coastline of California, a fog rolling in off the Pacific Ocean is not unusual. But during the summer months, people living in the southern part of the state, especially around the Los Angeles area, have often complained about burning eyes and have had difficulty breathing because of the smog—or acid fog. In a few instances the southern California fog has had a pH measurement equal to that of battery acid—hardly the kind of air one wants to breathe for good health.

Health hazards due to acid fog, rain, snow, or other precipitation are some of the most difficult to determine. A report from the U.S. Office of Technology Assessment indicated that the elderly and those with chronic heart and lung diseases could be at risk from acid fallout. Although OTA explains that the 30 million tons of sulfur dioxide emitted each year in the United States is not a health hazard "at normal environmental concentrations," the transformation products—sulfate particles—are of "greater concern."

According to the OTA, the tiny sulfate particles

> can readily be inhaled into the deep passages of the lung [and] acute exposures to sulfates . . . constrict lung passages and lengthen lung clearance times in humans and laboratory animals; chronic exposure of laboratory animals to sulfuric acid mist produced evidence of the onset of chronic lung disease.

In addition the report points out that numerous studies have shown "correlations between ambient sulfate concentrations and mortality rates." An estimated 51,000 premature deaths per year have been related to sulfur dioxide emissions. Deaths from respiratory diseases such as asthma and bronchitis could go much higher—to 57,000 premature deaths annually—if sulfur emissions are not curbed by the year 2000.

LOSSES IN MATERIALS

It is not just acid rain that causes damage to various structures—bridges, monuments, statues, paint finishes, etc. "It is dirty air that we are concerned with . . . [a] set of pollutants that we have to consider," according to Dr. Rudy Husar of Washington University. Advising a congressional committee on the effects of acidic deposition, Dr. Husar pointed out: "The national corrosion cost is estimated to be about five percent of the gross national product. A fraction of that is atmospheric corrosion and deterioration of materials. . . . " He believes there should be a cost-benefit analysis of damage to materials caused by "the entire system" of pollutants.

Research to determine the effects of acid precipitation on materials in the atmosphere, water, and soil has top priority in several government departments. Some of the research goals include learning what mechanisms are involved in materials damage, what differences there might be between the direct effects of acid rain and the effects of sulfur dioxide and nitrogen oxides that maybe converted to acid on the surface of materials.

The National Bureau of Standards will be developing methods to estimate costs for acid precipitation damage to original materials and for replacement. Maintenance costs and other expenses related to preserving materials also have to be estimated. In addition, the NBS will make predictions about how long certain materials will last.

Private industry will no doubt continue to develop coatings and treatments to protect a variety of materials from the effects of acid precipitation. But the federal government has the major responsibility for preserving historic structures such as Lincoln Memorial, the Statue of Liberty and the Washington Monument, which have already shown signs of deterioration thought to be caused by acid pollution. Laboratory and field programs are being established, and cooperative research with other nations is underway to protect historical artifacts—many in Europe have been more severely affected than those in the United States.

BILLIONS FOR SULFUR DIOXIDE CONTROLS

Whatever protective measures are taken, none can be more effective than stopping acid-producing substances at their source. While most controls do not apply directly to acid precipitation, controls now in effect do limit to some degree emissions of sulfur dioxide and nitrogen oxides—the forerunners of acidic substances—and standards for air quality are regulated by the Clean Air Act. To meet the clean air standards, various technologies are currently in use.

One example is a flue gas desulfurization system (called scrubbing) which controls pollution from some coal-fired power plants. With "wet scrubbing" jet streams of wet lime blast fumes in smokestacks, reducing from 70 to 95 percent of the sulfur dioxide before it gets into the atmosphere. Installing such a system on a *new* plant is considered relatively inexpensive in comparison to building such "retrofit" systems for all older plants in the nation. At $100 million apiece, in mid-1980 dollars, total initial outlay for utilities to control emissions with scrubbers has been estimated at $7 to $14 billion, with yearly operating costs totaling about $2 billion or more.

Another type of control is less costly and involves grinding up and washing coal before it leaves the mine. This process can remove much of the sulfur, and this, in turn, reduces sulfur dioxide emissions from coal-burning power plants. Sulfur can also be removed from coal by a chemical process, but this is much more expensive than the physical "coal washing."

Using low-sulfur coal would be one more way to cut back on sulfur dioxide pollutants. However, low-sulfur coal is more expensive to use since it must be transported from the West or from Kentucky and West Virginia mines to high consumption areas in the Midwest. Also, emission control devices at power plants would have to be modified in order to burn low-sulfur coal.

If power companies are required by law to switch to low-sulfur coal, "profound shifts will occur in coal production," says EPA spokesperson Kathleen Bennett. She predicts this would "impose large costs—in terms of jobs, coal market impacts, and consumer utility rates" in traditional coal-producing regions such as northern Appalachia and the Midwest. Ms. Bennett sees high unemployment among miners and shifts in other types of support jobs.

This is the view often expressed by the National Coal Association, which is against the Senate committee's proposal to reduce emissions by 8 million tons a year in a twelve-year program. Carl Bagge, president of the association, says that reducing emissions further means sacrificing production. He recently described this as a "perverse kind of progress" and concluded that "the coal industry and the American people will never support clean air progress through economic stagnation."

The EPA projects that the Senate committee's approved proposal for reducing emissions could cost $5 to $6 billion per year based on 1982 dollars.

> The cost per ton of sulfur dioxide removed would average around $500 . . . but in some states . . . would be greater than $1200 per ton removed. The average increase in annualized electricity rates in the 31 state region would be more than 4%, with some states experiencing increases greater than 7% [Ms. Bennett says].

NITROGEN OXIDES CONTROLS

Controlling nitrogen oxides emissions is accomplished by using low nitrogen oxides burners which include staged combustion processes. As OTA explains it:

> During the first stage of combustion, less air is supplied to the burner than is required to completely burn the fuel. Fuel-bound nitrogen is then released, but is converted to free nitrogen because oxidizing agents are not available. The subsequent addition of air causes the remaining fuel to be burned.

OTA states that nitrogen oxides emissions are reduced by this method "at a relatively low cost." The amount of nitrogen oxides reduction "depends on the type of fuel burned, the type of boiler in use, and whether the plant is new or already exists." The low-nitrogen oxides burner technique can reduce emissions in a coal-fired plant by 50 percent and 60 to 80 percent in an oil-fired plant, the OTA report notes.

CONTROL ISSUES: POLITICAL AND ECONOMIC

Whatever the amount of reduction in sulfur oxides and nitrogen oxides emissions required and the types of control technology used, there are and will be both political and economic impacts. Suppose legislation is passed to reduce sulfur dioxide emissions by 10 million tons in a thirty-one-state region east of the Mississippi River, as proposed by Senator Mitchell's bill (S. 1706). This would involve a combined political and economic decision as to who should reduce emissions and pay the costs. The political and economic issues are of concern to many segments of the population in that region.

For example, the proposal to bring western coal into the Midwest (as a way to alleviate sulfur oxides emissions) is politically explosive. The coal industry in the region is protesting and wants federal and state officials to protect area jobs and businesses. Some consumer groups are also protesting. Hikes in electric utility rates brought on by acid rain controls are not popular.

Yet increases in electricity prices might not be as great a burden to consumers as some have stated. The OTA report shows that several states which would be required to bring about the largest reductions in emissions, would still have some of the lowest electricity rates in the United States.

Another political/economic aspect of emission controls is the fact that electric rates are set by government regulatory agencies or commissions. Costs for reducing emissions would depend on how a regulatory agency allowed power companies to pass on expenses. Sometimes there are "automatic fuel adjustments." The costs for increased fuel prices are immediately added on to electric bills. In other cases, when utilities invest capital/money to install equipment such as scrubbers, regulatory agencies usually require that the control devices be operating before utilities can charge consumers for their extra costs. If a company cannot pass on increased costs to consumers right away, the management might decide not to invest in pollution control equipment and instead use a different type but more expensive low-sulfur fuel to meet control regulations.

There are a few state commissions, though, which allow costs for construction work in progress (CWIP) to be included as part of the base rate for electricity. The electric rates are increased over the short term, but over the long run or lifetime of the power plant, costs for producing electricity are decreased, and, theoretically at least, rates remain fairly stable.

A Senate bill has been proposed to help states which must bear the costs of emission controls. If certain states (and utilities) are required to adopt expensive control measures, the bill proposes that costs be shared by electric consumers in other areas. This could be done through a trust fund. Each utility in the region would contribute a fee, based on the number of kilowatt hours sold, to an "Acid Deposition Reduction Trust Fund." Money could then be allocated to states that are required by law to reduce emissions.

INVEST NOW OR LATER?

The legal counsel for the American Electric Power Service Corporation, a utility serving seven states in the Midwest, has called the debate about acid rain controls a "facade" and "simply a political device." The attorney contends the debate is "being used to . . . mask a predetermined objective—increased emission reductions" without any "causal linkage" between air pollution and acidification. The power company representative believes the impact of acid rain effects on the environment are mere hypotheses and "no basis for the initiation of corrective action."

On the other side of the issue, a number of scientists believe additional emission controls are needed now and are worth the cost. They insist some inexpensive initial steps should be taken. For one, state regulatory agencies could require electric utilities to operate their cleanest-burning plants for bulk loads; older, dirtier plants should be used only during times when electricity demands are at their highest. Coal-washing, as described before, is another nonstructural means of cutting back on sulfur dioxide.

Research scientist Orie Loucks believes such first steps must be followed by a program that would over the long term result in a 75 percent decrease in emissions. His studies have shown acid rain inputs would have to be reduced 75 percent in order to reach a level where there is no effect on the environment.

"Of course, it is improbable that we are going to do that very soon," Dr. Loucks says, but he thinks the 75 percent reduction can be "attacked incrementally, in three parts . . . . "

I like to use the analogy of someone with a severe cold, who has three long-term cold pills to take for relief, but is concerned about the side effects of taking each one. In emissions reduction we might have relatively little difficulty and only mild uncertainty about taking the first "pill" or step which might be a 20 to 25 percent reduction. We have more uncertainty about taking the second step and would like to fine-tune the cost-benefit relationship carefully before taking the third one.

The economics of the acid rain problem will be discussed and debated for some time to come. If, for example, as one scientist has suggested, it is proven that forests are threatened by acidic deposition, there could be a major impact on the U.S. economy. Paper and lumber companies need the forestry products, and such industries would support legislation to control emissions that cause acid pollution and losses in essential natural resources. In other areas of the economy, similar pressures could be felt. Thus, economic losses may eventually bring about some type of solution to the acidity that threatens parts of the environment and causes damage to materials.

CHAPTER 9

DETECTIVES ON WATER, LAND, AND IN THE SKY

What kind of scientific detective work is going on to solve the mysteries of acid rain? Who is doing what, and where? Will enough evidence be uncovered to pin down the culprits and stop the menace of acidic pollution?

LAKE 223

For nearly a decade Canadian scientists have been conducting investigations of acidic deposition in an Experimental Lakes Area near Kenora, Ontario. Set aside by the Canadian government in 1969, the area includes forty-six lakes which, like thousands of other lakes in eastern Canada and upstate New York, have a low-buffering granite underlay. All of the experimental lakes are far from sources of pollution, all have similar characteristics (including chemistry), and all are excellent experimental laboratories.

At first scientists studied phosphate inputs in one of the lakes. Such research helped bring about legislation to reduce the use of phosphate detergents which threatened Canadian and U.S. lakes and streams. By the mid-1970s

Canadian researchers were concentrating their efforts on studies of acid deposition effects on aquatic ecosystems.

A 50-acre lake—simply labeled 223—in the Experimental Lakes Area was chosen for acid precipitation experiments. No human activities polluted the watershed, and the pH of the lake in 1976 was 6.6. Since then scientists have been adding large amounts of sulfuric acid to Lake 223 and have been studying the effects.

During the first year the lake had natural buffering ability. There were no noticeable changes in pH or in aquatic life. From 1977 on, with continual acidic inputs, the pH has dropped to about 5, and there have been dramatic changes in the lake's chemistry and biology. Toxic metals have increased, leached from the lake sediments. The population of crayfish (a food for trout) has declined, and white suckers are not reproducing at a normal rate.

The primary food supply for trout—minnows and shrimps—are gone, although trout have not decreased in number. The scientists do not plan to lower the pH to kill off such sport fish. They already have enough evidence to conclude that pH levels below 5 destroy fish populations. The main purpose of the Lake 223 experiments is to determine early effects of acidic deposition and what stresses are placed on the aquatic ecosystem.

EPRI RESEARCH

A well-publicized lake study in the United States is being sponsored by the Electric Power Research Institute (EPRI), one of the largest private research groups. Called the Integrated Lake Watershed Acidification Study, the highly respected project is designed not only to determine water quality of three neighboring lakes in the Adirondacks (as described in chapter 3) but also to investigate acid precipitation effects on treetops, the chemistry of "throughfall," snow depth chemistry, and studies of soils. Scientists at Brookhaven National Laboratory, the U.S. Geological Survey, and six universities are involved in the EPRI-funded work.

Dr. Robert W. Brockson, who heads ecological studies for the institute, says "concern about the issue of acid rain first began to arise at EPRI in 1975. This was a considerable time before the press and public in the United States and Canada became aware of acid rain as a potential problem."

Supported by the electric utility industry, EPRI has spent millions each year since 1977 in research. Through the ecological studies program and two other EPRI programs—environmental physics and chemistry and environmental risk and issue analysis—scientists are studying pollutants or agents associated with electric power production and use. EPRI hopes to "determine actual effects on human beings and ecosystems and, ultimately, to use this information to determine whether additional control is necessary," Dr. Brockson recently explained to a congressional committee.

CLOUD DETECTIVES

A number of cloud chemistry experiments have been conducted here and abroad. In one effort EPRI has sponsored a joint program with the Central Electric Research Laboratories in Britain to determine chemical reactions in the atmosphere which lead to acid rain. Scientists, airborne in a specially equipped plane flying over the North Sea, collected water samples directly from clouds. The data will help explain how British air pollution might contribute to the formation of acid rain falling on Norway.

During the summer of 1980, under the direction of the Environmental Protection Agency, a dozen planes were equipped with monitoring instruments or other devices to track and measure, over a five-week period, a variety of air pollutants, including levels of ozone, sulfur oxides, nitrogen oxides, and hydrocarbons. One of the planes, a Cessna called

Chem One, carried writer Jon R. Luoma, who described his experience for *Audubon* Magazine:

> Flying through a sea of blue summer haze 4,000 feet above a green patchwork of central Ohio farmland, we were looking for a balloon . . . which atmospheric-chemistry researchers had released the previous day near the Conesville powerplant. The balloon, tracked by Federal Aviation Administration radar, was a sort of airborne tag, floating along in a cell of stuffy summer air associated with a high-pressure system that had inched into Ohio.

The pilot of *Chem One* flew the plane in a grid pattern at different altitudes around the balloon while instruments sampled the air coming in through pipes placed in the fuselage. On the ground at Ohio State University (in Columbus), other researchers followed the progress of the balloon as it traveled in a clockwise circle around the high-pressure cell.

According to Dr. Kenneth L. Demerjian, EPA's director of its Meteorology and Assessment Division, EPA has been involved in research relating to atmospheric processes in acid deposition since 1979. Recently he noted that funding for the program is over $4 million a year.

> To date, the program has supported large amounts of fieldwork on the chemistry of atmospheric formation of sulfates and nitrates, on vertical and horizontal transport mechanisms, and on wet scavenging and dry removal of acidic compounds. Despite the fact that this research has revealed some important findings, we need to understand a great deal more before results of this research can serve as the basis for regulatory decisions [Dr. Demerjian said].

TACKLING TOMBSTONES

Researchers need to analyze uniform materials when studying damages to stone caused by acid deposition. Ideally, the materials should be produced under controlled conditions and be exposed to a variety of climates. Placed in different environments, the materials could then be examined and analyzed at certain times over a continuous time frame.

Such a study is being sponsored by EPA, using the marble headstones and markers in Veterans Administration cemeteries. The tombstones are fairly standard in size and shape and materials. Cemeteries in three climate zones were selected for the research. According to an EPA report, the stone markers are being examined "for such effects as measurable loss of detail, rounding of edges, and surface erosion. . . . This damage will then be correlated with data on the stones' history from Veterans Administration records and data on air pollution and meteorological patterns from the National Weather Service."

COOPERATIVE EFFORTS

To assess the current knowledge of acid rain effects, five joint U.S.-Canadian research groups have been at work. Other investigations are underway by provincial governments in Canada and state governments in the United States. In Wisconsin, for example, a review committee has been set up with representatives from industry and government and citizen task forces on acid rain. Wisconsin's Joint Acid Deposition Technical Review Committee hopes to integrate various research efforts and come up with an overall research plan for the state.

On a federal level the National Acid Precipitation Assessment Plan calls for a dozen federal departments and agencies to conduct some forty research tasks through 1992. These tasks include cooperative efforts with universities, and ongoing projects or programs include

- identifying the sources of atmospheric emissions which contribute to acid rain;
- monitoring and measuring through NADP levels of acid precipitation nationwide;
- conducting research in atmospheric physics and chemistry to help explain the processes by which substances in the atmosphere are converted to acids;
- developing models to help predict long-range transport of acid-producing substances;
- defining geographic areas at risk or sensitive to acid rain;
- broadening the data base on water and soil chemistry and analyzing trends of acid rain effects;
- studying the dose-response relationship (acidic deposits and effects) for various ecosystems in the environment;
- assesssing the economic effects of acid precipitation and remedies for possible harmful effects;
- coordinating various federal activities in regard to acid rain to avoid duplication and waste;
- developing cooperation between affected areas and those areas contributing to acid rain effects;
- analyzing information on acid precipitation in order to present recommendations for alleviating acid precipitation and its effects.

A total of over $40 million was budgeted for federal acid rain research for 1982 and 1983. But the funds represent only part of the combined interagency efforts. The National Science Foundation, for example, supports extensive research, but NSF activities are not included in the budget for the National Acid Precipitation Program. Even so, NSF studies are coordinated with the national program, and NSF representatives are part of the Task Force on Acid Precipitation. Similar cooperative efforts are taking place with other types of acid rain research, particularly in the development of control technology

CHAPTER 10

WHAT CITIZENS CAN DO

"Who will stop the acid rain?" is the question posed on a poster distributed by the Canadian Coalition on Acid Rain. Since its formation in December 1980, the Coalition has been pressing governments in both Canada and the United States for solutions to the acid pollution problem. Composed of representatives from forty member groups with business, conservation, environmental, and recreational interests, the organization has been campaigning for a regulatory program on acid rain to be included as part of the U.S. Clean Air Act. The citizen group also is urging that the best available control technology be installed in all point sources of sulfur dioxide and nitrogen oxides in Canada and would like to see a 50 percent reduction in sulfur dioxide and nitrogen oxides emissions by the year 2000.

Another major effort of the Canadian coalition is educating Americans on what effect acid rain originating in the United States is having on eastern Canada. A variety of materials is available from the coalition, including a well-documented "Fact Sheet on Acid Rain" and a booklet of articles and editorials on acid rain from the *Cleveland Plain*

Dealer. Titled "Poison from the Skies?" the booklet contains features which clearly point up the political aspects of the acid rain issues and explains the need for effective regulations to reduce acid-producing substances.

The coalition welcomes inquiries, and you can obtain information by writing: Canadian Coalition on Acid Rain, 112 St. Clair Avenue West, Suite 504, Toronto, Ontario M4V 2Y3. Or you can call (416) 968-2135.

U.S. EFFORTS

The counterpart of the Canadian public interest group in the United States is the National Clean Air Coalition. With headquarters at 530 7th Street, S.E., Washington, DC 20036 (Phone: 202/543-8200), the organization is involved in lobbying efforts to strengthen the Clean Air Act.

Another organization working to solve the acid rain puzzle is the Acid Rain Foundation, 1630 Blackhawk Hills, St. Paul, MN 55122 (Phone: 612/454-2621). Incorporated in December 1981, the foundation officers include scientists such as Ellis Cowling of NCSU and Eville Gorham of the University of Minnesota who have been in the forefront of acid rain research.

What does the foundation do? Some specific services include sending out information on acid rain to the media, interested groups, and individuals. The foundation encourages and helps coordinate meetings, workshops, and conferences on acid rain and has organized a speakers' bureau to create public awareness of the pollution problem. Curricular materials for teachers and students, fellowships, and research support are other services the foundation provides.

Many other national organizations are concerned about the effects of acid rain. Some conservation and environmental groups have been involved for a long time in efforts to protect clean air, forestry, and wilderness areas, and to conserve energy and natural resources. You can learn more about their programs to combat acid rain by contacting the national headquarters for groups such as these:

National Audubon Society
950 Third Avenue
New York, NY 10022

National Wildlife Federation
1412 16th Street N.W.
Washington, DC 20036

Sierra Club
530 Bush Street
San Francisco, CA 94108

Soil Conservation Society of America
7515 N.E. Ankeny Road
Ankeny, IA 50021

Also, check directories of national associations. A number of new organizations have developed in the past few years to lobby against federal efforts to relax regulations which protect the environment.

GRASSROOTS GROUP

In the spring and summer of 1982, nearly 140 volunteers in different sites scattered across the state of Vermont armed themselves with plastic buckets. Part of an acid rain monitoring effort sponsored by the Lake Champlain Committee, a conservation group, the citizens were ready to wage their collective battle. They nailed their one-quart containers to posts or tree stumps in the wide-open spaces of backyards or vacant areas near places of business. Then they waited for the raindrops to fall.

During rain events the volunteers collected samples,

then tested the water for acidity, using inexpensive litmus paper which measured the pH. Along with other information about each rain event, pH measurements were recorded on an "Acid Precipitation Data Sheet." Twice a month volunteers, who paid three dollars each to cover costs of their acid rain monitoring kits, sent their logged information to the Lake Champlain Committee, which is under the direction of Executive Secretary Anne Baker.

The Lake Champlain Committee is working with the botany and geography departments of the University of Vermont where data is being analyzed by computer and translated onto maps. Precipitation data and physical features of the Lake Champlain Drainage Basin will be useful to government officials, scientists, and other acid rain researchers.

Extending west from the Adirondack Mountains in New York to the Green Mountains in Vermont on the east, and from Quebec in the north to the New York–Vermont borders in the south, "the Lake Champlain Basin is an ideal unit for an intensive monitoring study of acidic precipitation," said Dr. Richard Klein of the botany department at the University of Vermont. "There is an unusually complete data base of the geology, physical geography and water resources of the region," he noted, adding that it is possible to locate potential sources of pollution precisely and to measure known meteorological patterns.

Data collected by the LCC volunteers has resulted in analysis maps such as the first one released in the fall of 1982. The acid precipitation readings for June, as shown, indicate a range of 4.25 to 4.5 pH for 84 percent of the 8,234 square-mile basin—over ten times more acidic than so-called "clean" rain. Some areas showed an even lower pH range of 4 to 4.25.

Executive Secretary Anne Baker of LCC cautioned that many years of data are required to "establish long range trends." But already the pH readings suggest that the atmosphere is saturated with sulfur and nitrogen oxides, according to Dr. Hubert Vogelmann, chairman of the botany department at the University of Vermont.

SCHOOL PROJECTS

Independent School District #197 of West St. Paul, Minnesota, has been involved for several years in a project to create Acid Precipitation Awareness materials for junior and senior high schools. Directed by Harriett A. Stubbs, the Title IV-funded program has been developed by teachers, science and education consultants, and environmental and natural resource experts. A number of acid rain packets have been published, and they include activities to help students in grades seven to twelve "learn science while also learning about a major environmental problem: acid precipitation."

Have you ever wanted to observe how acid affects materials? One activity helps students test the effects of acidity (prepared in the chemistry laboratory with sulfuric acid) on various kinds of fabrics. In such experiments, logs are kept showing the type of damage to, say, nylon after applying drops of solution ranging from a pH 1 to 6.4. Students also learn to draw conclusions about possible effects of acid rain and the variables measured in the experiments.

During another activity students collect and analyze snow for acidity. They learn how to select sites for samples and compare pH levels of different types of snow such as samples of fresh snow, snow several days old, and snow near and far from traffic.

If you are interested in acid rain experiments or would like your school class to get involved in such activities, the learning materials are available at a nominal cost. Write to the Acid Precipitation Awareness project, 1037 Bidwell Street, West St. Paul, MN 55118. Or you can call (612) 455-7719.

WHAT CAN ONE PERSON DO?

With a problem as complex and political as acidic pollution, it often seems that the actions of one person make little difference. But when many individuals have joined together to support a cause (or to protest), they have often been effective in bringing about change. That is what coalitions are all about. Yet even alone, one person *can* do something—or some things.

Write letters. It is one of the first steps for many action groups. You have probably read the suggestion or heard someone urge: "Write your congressional representative." Sometimes this can be a class project in which students express their concerns through letters to members of Congress. You can also write to the president of the United States and can make known your opinions about acid rain to those who head congressional committees or federal agencies investigating the effects and control of acid rain.

Get information. This chapter includes the names and addresses of several private organizations which distribute a great deal of information about the effects of acid rain. You can also request materials from other public interest groups and government agencies such as the following:

The Environmental Defense Fund
1525 Eighteenth Street N.W.
Washington, DC 20036

Environmental Law Institute
1346 Connecticut Avenue N.W.
Washington, DC 20036

Pollution Probe
12 Madison Avenue
Toronto, Ontario M5R 2S1
Canada

United States
Environmental Protection Agency
Washington, DC 20460

Interagency Task Force on
Acid Precipitation
722 Jackson Place, N.W.
Washington, DC 20006

Environmental Protection Service
Environment Canada
Place Vincent Massey
Ottawa, Ontario K1A 1C8
Canada

Contact power companies, smelters, and similar industrial groups/companies to learn what they are doing about acid rain controls.
For a start, you can write to

American Electric Power Service Corporation
180 East Broad Street
Columbus, Ohio 43215

Edison Electric Institute
1111 19th Street, N.W.
Washington, DC 20036

EPRI Public Information Department
P.O.Box 10412
Palo Alto, CA 94303

National Council of the Paper Industry
for Air and Stream Improvement, Inc.
260 Madison Avenue
New York, NY 10016

INCO Limited
P.O. Box 44
1 First Canadian Place
Toronto, Ontario M5X 1C4
Canada

Ontario Hydro
700 University Avenue
Toronto, Ontario M5G 1X6
Canada

Plan an awareness campaign. This could be a class or school project or part of a science fair. You or members of your family might belong to a civic or service organization willing to support efforts to inform the public about acidic pollution. Again, materials and ideas are available from acid rain coalitions and other groups just described. Bumper stickers and posters help inform others. You might create your own.

Get involved. Citizen action groups always need volunteers. Check with science teachers, environmental groups, and others in your area to see if some kind of action program in regard to acid rain is underway or being planned. Offer your services. You are bound to be welcome.

Save energy. You've heard or seen that slogan before. But conservation is always important. There are countless publications on how to save energy. You can learn the types of appliances, heating, insulation, transportation, and so on that conserve or cut down on the use of fossil fuels and thus reduce emissions which cause acidic pollutants. Serious efforts to make better use of our energy resources can go a long way toward protecting our environment from the impacts of acidic deposition.

FOR FURTHER READING

Bass, Thomas. "Acid Rain: Deadly Fallout Perils Remote Watersheds in Rockies." *Audubon*, November 1982, pp. 116–18.

Boyle, Robert and Boyle, Alexander R. *Acid Rain*. New York: Schocken, 1983

Boyle, Robert. "American Tragedy," *Sports Illustrated*, September 21, 1981, pp. 70–82.

Eckholm, Erik P. *Down to Earth: Environment and Human Needs* (Chapter 7). New York: W.W. Norton & Company, 1982.

Ember, Lois R. "Acid Pollutants: Hitchhikers Ride the Wind," *Chemical and Engineering News*, September 14, 1981, pp. 20–31.

Howard, Ross, and Perley, Michael. *Acid Rain* (paperback edition). New York: McGraw-Hill Book Company, 1982.

Hoyle, Russ. "The Silent Scourge," *Time*, November 8, 1982, pp. 98–104.

LaBastille, Anne. "Acid Rain—How Great a Menace?" *National Geographic*, November 1981, pp. 653–80.

Likens, Gene E., and others. "Acid Rain," *Scientific American*, October 1979, pp. 43–51.

Ostmann, Robert, Jr. *Acid Rain—A Plague upon the Waters*. Minneapolis, Minn.: Dillon Press, 1982.

U.S. Interagency Task Force on Acid Precipitation. *National Acid Precipitation Assessment Plan*. Washington, D.C.: U.S. Government Printing Office, 1982.

Vogelmann, Hubert W. "Catastrophe on Camel's Hump," *Natural History*, November 1982, pp. 8–14.

INDEX

"Acid-altered" changes, 20
Acid deposition, 3
Acidity, increase in, 4–7
    Interagency Task Force on Acid Precipitation, 5
    monitoring networks, 5–6
    pH in rainfall, U.S., 4–5
    sources of, 6–7
Acidity, measurement of, 3–4
Acid precipitation, 3
Acid Precipitation Awareness project, Minnesota, 76
Acid rain, 2
Acid rain bills, 46–47
Acid rain controversy
    Canada, 12–13
    historical studies, 10–11
    media, 9, 38
    modern research, 11–12
    "scare word," 38–39
    U.S., 13–14
Acid Rain Coordinating Committee, 44
Acid Rain Foundation, 73
Acid rain, impact on environment, 8, 14
"Acid shock," 18
"Acid test," 38
Algae and pH, 17–18
Alkalinity levels, water, 20
American Civil Liberties Union, 50–51
American Electric Power Service Corporation, 63, 78
American Electric Power System, 38
Amphibians and pH, 17
Anions, 3
Asbestos, 22
Atmosphere, 21
Automobiles, paint finishes, 2

Bagge, Carl, 60
Baker, Anne, 75
Bennett, Kathleen, 47, 60
Big Moose Lake, 15
Boundary Waters Canoe area, Minnesota, 5, 15

Brockson, Robert W., 41, 67
Brookhaven National Laboratory, 66–67
"Buffering" capacity of waters, 16–17

California, 54, 57
   Assembly Select Committee on Acid Rain, 54
Camel's Hump, 24
Canada
   acid rain studies, 12–13
   films, 50
   fish industry, 56
   utilities, 41–43
   versus U.S. controversy, 43, 47–49
Canadian Coalition on Acid Rain, 42, 72, 73
Canadian Network for Sampling Precipitation Program (CANSAP), 6
Carter, President Jimmy, 44
Cations, 3
Central Electric Research Laboratories, Great Britain, 67
"Chemical fallout," 2
Citizen's program for action, 77–79
   awareness campaign, 79
   get involved, 79
   information sources, 77–79
   letters, 77
   save energy, 79
Clean Air Act. *See* U.S. Clean Air Act
"Clean" rainfall, 4
Coalition for Environmental Energy Balance (CEEB), 37
Coal versus oil, 39
Cogbill, Charles, 13
Conference on Acidification of the Environment (1982), 52–53

Construction work in progress (CWIP), 62
Control rain, 27
Corvallis Agricultural Experiment Station, Oregon State University, 27–28
Costs to environment, measurement of, 55–64
   estimating damages, 55–57
   health costs, 57–58
   losses in materials, 58–59
   nitrogen oxides controls, 61
   political and economic control issues, 61–63
   program steps, 63–64
   sulfur dioxide controls, 59–60
Cowling, Ellis B., 6, 13, 14, 51, 73
Crops, marketability of, 27

Demerjian, Kenneth L., 68
Drinking water quality, 21–22
Dry deposition, 3, 5
Dustfall, 3

Economics of acid rain, 64. *See also* Costs
Edison Electric Institute, 38, 78
Edwards, James, 9
Electricity prices, increases, 62
Electric Power Research Institute (EPRI), 41, 78
   lake studies, 17
   research, 66–67
Elemental tracers, 34–35
Environmental Defense Fund, 77
Environmental Law Institute, 77
Environmental Protection Agency (EPA). *See* U.S. Environmental Protection Agency

Falun, Sweden, 10
Ferguson, W. Charles, 42
Fir trees, 2
Fish and pH, 18, 21

Flacke, Robert, 56
Flue gas desulfurization system, 59
Foreign Agents Registration Act, 50
Forest floor, 25
Forest products industry, 26
Fossil fuel combustion, 6

Galloway, James N., 13–14
Gaseous ozone, 3
General Motors, 39
Glass, Gary, 48–49
Great Britain, 30–31
Great Lakes Water Quality Agreement, 51–52
Green Mountains, Vermont, 23–24
Gorham, Eville, 11, 73

Harte, John, 19
Harvey, Harold, 12–13
Husar, Rudy, 58
Hydrogen ion concentration, 3

INCO Ltd., 42, 79
Integrated Lake Watershed Acidification Study, 66
Interagency Task Force on Acid Precipitation, 5, 44, 78
International cooperation, 52–53
Ions, 3

Klein, Richard, 75

Lake 223, 65–66
Lake Champlain Committee, 74–75
Land and sea, 30–35
  Arctic, 34
  elemental tracers, 34–35
  remote areas, 31–32
  state and national boundaries, 32–33

Lead, 21, 22
Legislation against acid rain, 44–54
  acid rain bills, 46–47
  Canadian–U.S. treaty problems, 47–49
  Clean Air Act, 45–46
  Great Lakes Water Quality Agreement, 50–51
  international cooperation, 52–53
  propaganda, 50–51
  state actions, 53–54
Likens, Gene E., 13
Liming, 20
Limnology, 11
"Loading Limit," 48
Loucks, Orie L., 16, 37–38, 63–64
Luoma, Jon R., 68
Lugar, Richard, 36
Lyons, Walt, 32–33

Megonnell, William, 38
Mercury, release of, 18
Minnesota, 54
Mitchell, George, 46, 61

National Academy of Sciences, Research Council, 40
National Acid Precipitation Assessment Plan, 45, 69–70
National Audubon Society, 74
National Bureau of Standards, 58
National Clean Air Coalition, 73
National Council of the Paper Industry for Air and Stream Improvement, Inc., 78
National Science Foundation, 70
National Wildlife Federation, 74
Nelson Lake, Ontario, 20
Nitrogen deposits, fertilizing benefits, 26
Nitrogen fixation, 27
Nitrogen oxides (NOx), 6–7, 61

Oden, Svante, 11–12
Office of Technology Assessment (OTA)
    algae community structure, 18
    eastern U.S. lakes and streams, 19
    forest benefits, 26
    nitrogen oxides control, 61
Ohio River Basin, 7, 32
Oil versus coal, 39–40
Ontario Hydro, 79
Ozone effects, plants, 28–29

Panther Lake, 41
Perley, Michael, 42
pH (potential hydrogen), 3, 17–18
    scale, 4
Pipes and conduits, 21
Pollution Probe, 77
Precursors, acid rain, 6–7
Public Service of Indiana, 37

Rahn, Kenneth A., 34–35
Reagan, President Ronald, 44
Rocky Mountains, Colorado, 19

Sagamore Lake, 41
Schofield, Carl, 13
Scrubbing, 59
Sierra Club, 74
Smith, Robert Angus, 10–11
Soil Conservation of America, 74
Soil management, 27
Stafford, Robert, 46
Stubbs, Harriett A., 76
Sulfur dioxide controls, 59–60
Sulfur oxides (SOx), 6–7
Sweden, 2, 30

Tall stacks, 8
    EPA limit, 45
Task Force on Acid Precipitation, 71
Transboundary air pollution, 52
Trees and crops, 23–29
    benefits, 26–27
    crops, effects on, 27–28
    decreasing growth rates, 1
    forest studies, 24–26
    ozone effects, 28–29
    questions asked, 29

U.S. acid rain studies, 13–14
U.S.–Canada
    airshed, 32
    controversy, 43, 47–49
    research groups, 69–71
U.S. Clean Air Act, 42, 72
    amendents, 36, 45, 47
    EPA, 45
U.S. Environmental Protection Agency (EPA), 78
    Information Services, 2–3
    reduced-emission costs, 60
    studies, 67–69

Vogelmann, Hubert W., 1–2, 24, 76

Waters, condition of, 15–22
    acidified waters, 17–19
    affected waters, number of, 19–20
    buffering capacity, 16–17
    countermeasures, 20–21
    critically acidified lakes, 15
    drinking water quality, 21–22
    physical description, 16
    pure, 16
Watershed, 21
Webster, Dwight, 13
West Germany, forest studies, 24–25
Wet deposition, 3
Whitehead, Donald, 32
Wind and acid-forming materials, 7–8
Wisconsin, 54, 69
Woods Lake, 15, 41

Zooplankton, 18